Basic Studies in Environmental Knowledge, Technology, Evaluation, and Strategy

Takayuki Shimaoka • Takahiro Kuba •
Hirofumi Nakayama • Toshiyuki Fujita •
Nobuhiro Horii
Editors

Basic Studies in Environmental Knowledge, Technology, Evaluation, and Strategy

Introduction to East Asia Environmental Studies

Springer

Editors
Takayuki Shimaoka
Kyushu University
Fukuoka, Japan

Takahiro Kuba
Kyushu University
Fukuoka, Japan

Hirofumi Nakayama
Kyushu University
Fukuoka, Japan

Toshiyuki Fujita
Kyushu University
Fukuoka, Japan

Nobuhiro Horii
Kyushu University
Fukuoka, Japan

ISBN 978-4-431-55817-0 ISBN 978-4-431-55819-4 (eBook)
DOI 10.1007/978-4-431-55819-4

Library of Congress Control Number: 2016931989

Springer Tokyo Heidelberg New York Dordrecht London

Printed on acid-free paper

Springer Japan KK is part of Springer Science+Business Media (www.springer.com)

Preface

Human activities today are increasing at a pace rarely seen in human history. The economies in the West and in developed countries in other parts of the world have continued to grow over the long term because of mass consumption and at the expense of resources and energy. Population growth, urbanization, and economic development likewise have progressed in developing countries, including those in East Asia, at a speed greater than that in the developed countries; consequently, developing countries have had to deal with environmental problems more seriously than developed countries have, including water and air pollution, soil erosion, deforestation, loss of biodiversity, and desertification.

A major characteristic of East Asia is a great discrepancy in income between the developed and the poorest countries, even as the region as a whole is growing, with the developing countries chasing after the developed countries as they step up the rate of their stages of development. East Asia is thriving in certain areas because of international relationships in production, trade, investment, finance, and aid.

Mutual dependency between Japan and East Asia deepened when the East Asian countries opened up their respective economies under various liberalization policies that promoted trade with Japan and acceptance of direct Japanese investment as keys to their economic development strategy. These deepened relationships have engendered high economic growth in this area.

Besides the existing economically mutually dependent relationships, a proposal for new coordination in environmental conservation as an international mission would be significant in terms of the current international situation. Therefore, as a developed country, Japan is expected to play a large role in East Asia regarding environmental problems. Japan possesses an environmental awareness born of its struggles to overcome pollution and its accumulation of rich technical knowledge, funding, human resources, and research for solving problems; and the significance of reinvesting such resources in foreign countries is enormous. International cooperation and coordination in environmental conservation in all of East Asia must be recognized as essential to the stability and development of the region, and the development of a system for fostering such relationships is absolutely necessary.

To solve the environmental problems in East Asia, the regional diversity in geography, climate, history, religion, culture, political state, and economy must be recognized. The environmental sense that is common to the region and that emphasizes harmony with the natural environment must be understood, and strategic solutions as per local characteristics and conditions must be assessed while taking the diversity and commonalities in the region into account. For example, accounting for factors such as local economic, technical, and human resource standards in policy designs and technical selections is important for developing countries. Therefore, it is absolutely necessary to train people who are capable of finding strategic solutions through wide, highly professional knowledge in not only environmental problems but also related fields (i.e., "environmental strategists," which we will discuss later), and to take an academic approach by systematizing East Asia Environmental Studies. Here, we define East Asia Environmental Studies as courses of studies for comprehensively understanding the essence of environmental phenomena in East Asia with the characteristics discussed earlier, and for designing optimal policies and technologies for creating a sustainable, safe, and secure environment.

Kyushu University has a long history in and extensive knowledge of international coordination with East Asian countries because of its geographical advantage, and it has engaged in many environmental studies with a strong awareness of the transboundary damage inflicted by environmental problems from China, the Korean Peninsula, and Southeast Asia. The East Asia Environmental Problem Project that began in September 2007 was a part of Kyushu University's 100th anniversary project, which addressed complex global environmental problems such as air pollution, river and ocean pollution, urban problems, wastes, and food pollution in East Asia, particularly in China, which is undergoing rapid development.

The project was reformed and expanded in April 2009 to the Research Institute for East Asia Environments (RIEAE) with the president of Kyushu University as the head, and it has developed a system for contributing to responses concerning environmental issues in East Asia with education and research.

RIEAE and its researchers aim for social contribution at a national level by solving increasingly complex environmental problems in a practical manner through coordination with relevant institutions. With support from private companies, RIEAE has established 10 research groups, as follows. The "social infrastructure consortium" (urban environment group, low-carbon urban system group, food risk group) works to solve environmental problems caused by urbanization and industrialization; the "environmental symbiosis consortium" (aquatic environment group, anti-desertification group, bioproduction environment group) conducts research for the sustainable and effective use of natural resources without loss of biodiversity; and the "environmental conservation consortium" (air quality group, marine environment group, and environmental chemistry group) and "environmental planning and policy group" research the prevention of environmental pollution at the local, regional, and trans-boundary levels. In all, 58 researchers are linked to one another and are working on diverse themes in research and education.

Kyushu University has positioned the RIEAE as one of its major projects, and it is funded by the Special Research Budget of the Ministry of Education, Culture, Sports, Science and Technology. The institute was founded initially for research purposes, but as stated earlier, human development is absolutely essential for solving environmental problems. Therefore, the institute began to work on developing human resources and providing the research results to education, which truly took off when the East Asia Environmental Strategist Training Program proposed by the institute was selected for the next 5 years in October 2010 by the Japan Science and Technology Agency (JST) as the International Environment Leaders Training Program.

The East Asia Environmental Strategist Training Program is for the purpose of training international leaders by playing to the strengths of Kyushu University as a leader in East Asian environmental studies. It employs the university's rich human resources and its educational and research facilities, as well as the geographical advantage of being near advanced research institutions such as RIEAE.

International leaders trained through the program are called "environmental strategists." The qualifications for an environmental strategist have been defined as comprising the four following skills:

- Environmental knowledge: has acquired a wide range of knowledge from the social to the natural sciences and can comprehensively and systematically understand the relationship between human activities and environmental problems from various points of view.
- Environmental technology: is able to choose feasible technologies from a menu of environmental measures based on local conditions in a developing country.
- Environmental evaluation: is able to accurately understand the structure of environmental problems, and is thoroughly knowledgeable about tools for assessing environmental load and impacts.
- Environmental strategy: is able to apply solutions with leadership using strategic thinking skills, as well as decision making for solving environmental problems in a consensus formation process.

The major goal of this book is to train environmental strategists, using the essence of the East Asia Environmental Strategist Training Program. Its ultimate objective is to impart a broad knowledge of the environmental problems in East Asia in simple terms so that beginners can comprehensively and systematically understand environmental problems, and to help readers acquire four practical skills for strategic development of solutions: (1) environmental knowledge, (2) environmental technology, (3) environmental evaluation, and (4) environmental strategy.

As mentioned earlier, East Asia Environmental Studies is designed to foster a comprehensive understanding of the essence of social and environmental phenomena in East Asia and to develop optimal policies and technologies for creating a sustainable safe and secure environment. Developing environmental strategists who can find strategic solutions by using their wide and highly professional knowledge on environmental problems and related fields is inseparable from East Asia Environmental Studies as a whole. Therefore, this book is titled *Basic Studies in Environmental Knowledge, Technology, Evaluation, and Strategy: Introduction to Asia Environmental Studies*, and the contents are divided into four essential parts that correspond to the above-named four practical skills required for environmental strategists.

In this book, the authors have attempted to keep the text simple and to limit the explanations to basic concepts in each related field so that the content is understandable both to beginners, such as college undergraduates, and to professionals from different fields. In addition, the book is aimed at providing up-to-date and useful information and research so that it will be useful to front-line researchers and engineers. Part of the book was written by experts on the latest results of their research. We, the editors, would like to thank them for their contributions.

We will be greatly pleased if the book is of any help in solving the environmental problems in East Asia and beyond.

Fukuoka, Japan Takayuki Shimaoka

Contents

Editors and Contributors

Takayuki Shimaoka Takayuki Shimaoka, is a Professor of Faculty of Engineering, Kyushu University. He has B. Eng., M. Eng. and Dr. Eng. from Kyushu University. His research fields are environmental engineering and solid waste management. His specific research topics are qualitative change of solid waste and transportation of materials in landfill. In recent years, he has focused on geochemical change of incineration residues for the beneficial reuse of it. He acted an executive director of Research Institute for East Asia Environments, Kyushu University since 2009, and serviced to development of human resources who can contribute to solve Asian environment issues.

Nobuhiro Horii is an Associate Professor at Faculty of Economics, Kyushu University. He has B.Law and M.law (major in politics) from Keio University. His research work has been about energy and environment issues, especially focusing on China. He has been worked as a researcher of Institute of Developing Economies (IDE-JETRO), as a consultant of the World Bank and International Energy Agency, and as a visiting professor of Institute of Social Science, The University of Tokyo.

Takahiro Kuba is the professor of Department of Urban and Environmental Engineering, Graduate School of Engineering, Kyushu University. He has B.Eng. from the Saga University, and M.Eng. and Ph.D. in Eng. from the Kyushu University. His researches have mainly related to the integrated preservation of the water environment using various technologies, such as water environment purification, integrated water management, and sewage treatment, from a wide range of viewpoints that cover the large scale of urban ecology and the small scale of microorganism ecology. His current and previous research interests are the application possibility of bamboo charcoal as cesium ion adsorbent, development of "Eco-engineering Dam Barrier Structure (EDBS)", anaerobic wastewater treatment in a fluidized bed reactor, sewage treatment with denitrifying phosphate removing organisms and nitrifiers, and allelopathic effects of macrophyte on growth of cyanobacteria *Microcystis aeruginosa etc.* He has worked as a postdoctoral fellow at the Delft University of Technology, the Netherlands, and as an associate professor at the Kyushu University.

Hirofumi Nakayama is an associate professor of Department of Urban and Environmental Engineering, Faculty of Engineering, Kyushu University, Japan. Dr. Nakayama's major research interests are in the areas of environmental accounting on waste management, application of remote sensing technology for waste management.

Toshiyuki Fujita is a Professor at Faculty of Economics, Kyushu University. He has B.Eng., M. Eng., and Ph.D. in Eng. from University of Tokyo. His major is environmental economics and his research work has been about game theoretic analyses of global environmental policies and international environmental agreements. He has worked at Tokyo University of Science as a research associate and Kyushu University as an associate professor.

Tetsuya Kusuda is Senior Advisor at the Research Institute of East Asia on Environment. He has earned B.Eng., M.Eng., and Dr. Eng. from Kyushu University. His research work has expanded to environment, environmental systems, ecological systems, water and wastewater treatment, trenchless technology on underground pipelines and environmental ethics. He has worked for Kyushu University as professor and director of Institute of Environmental Systems, the University of Kitakyushu as professor, and University of Florida, USA as visiting professor. He was a member of Science Council of Japan and is a member of the Engineering Academy of Japan.

Reiitsu Kojima is Emeritus professor of Daitobunka Univ.. He worked as a senior researcher, at Institute of Developing Economics, a professor, dean and trustee of Daitobunka Univ., and a visiting professor of Peking Univ. and Peking Univ. of Foreign Language.

Michikazu Kojima is a senior research fellow in the Institute of Developing Economies, Japan External Trade Organization. He got M.Sc. in Agricultural and Resource Economics from University of California, Berkeley. His recent works are International Trade in Recyclable and Hazardous Waste in Asia (ed. with Etsuyo Michida) Edward Elgar 2013, and Kojima, Michikazu Policy for Fostering "Sound" Recycling Industries, Institute of Developing Economies and Institute for Global Environmental Studies, 2014.

Akihisa Mori is an associate professor of Graduate School of Global Environmental Studies, Kyoto University, serves as a director and Secretary General of the East Asian Association of Environmental and Resource Economics. He has conducted research on environmental aid, climate finance, as well as energy, climate and environmental integration in the context of economic development in East Asia. He edited several books, including Environmental Governance for Sustainable Development: An East Asian Perspective, United Nations University Press, 2013, The Green Fiscal Mechanism and Reform for Low Carbon Development: East Asia and Europe, Routledge, 2013 (with P. Ekins et al.) and Green Growth and Low Carbon Development in East Asia, Routledge, 2013 (with F. Yoshida).

Yuko Ishibashi is a senior researcher at Fukuoka Institute of Health and Environmental Sciences belonging to Fukuoka prefecture. Her research work has been about water environmental science.

Yuka Watanabe is an Associate Professor at East Asia Environmental Strategist Training Program, Research Institute for East Asia Environments, Kyushu University. She has earned her bachelor's degree in Liberal Arts from International Christian University, Master's degree in Environmental Sciences from Tsukuba University, and Doctoral degree in Science from Kyushu University. Her research focuses on environmental chemistry and education. She has worked at the National Institute for Environmental Studies as a researcher, Kurume University, School of Medicine as an Assistant Professor, and Kyushu University Graduate school of Engineering as an Assistant Professor.

Hiroshi Kumagai is an administrative officer at Environmental Preservation Division, Department of Environmental Affairs, Fukuoka Prefectural Government. He has B.Eng., M.Eng., and Ph. D. in Eng. from Kyushu university. His research work has been about water environmental engineering. He had worked at Water Quality Division, Fukuoka Institute of Health and Environmental Sciences.

Hiroki Tanikawa is a Professor at the Graduate School of Environmental Studies at Nagoya University, Japan, and a Leader of Nagoya University Global Environmental Leaders Program (NUGELP). Since 1998, the Tanikawa Laboratory, which focused on the field of Environmental System Analysis, is estimating and evaluating the weight of human activity, and particularly attempting to elucidate how much material has been used in industrial processes.

Hideo Koide is Professor at Seinan Gakuin University. He has Bachelor of Economics from Yokohama National University, Master of Economics and Ph.D. in Economics from Hitotsubashi University. His research work has been about environmental economics, environmental policy management and design of interdisciplinary research and education. His doctoral thesis has been published as the title of "Resource Circulation Economy and Internalization of Externalities" from Keiso Shobo in November 2008.

Mitsuyasu Yabe is a Professor at the Department of Agricultural and Resource Economics at Kyushu University in Japan. His areas of expertise are agricultural economics and environmental economics. He has been involved in a variety of research activities in the economic evaluation on biodiversity conservation, rural amenity food safety, and biomass.

Shinji Kaneko is professor at Graduate School for International Development and Cooperation (IDEC), Hiroshima University since 2009. He received Dr. of Engineering from Kyushu University in 1999. He has worked for Institute for Global Environmental Strategies (IGES) based in Hayama, Japan from 1999 to 2002 as a researcher and has joined Hiroshima University in 2002 as an associate professor in charge of researches and education in development economics and environmental economics. His research achievements include approximately 40 publications in SCI journals and a book entitled Environmental Policy and Governance by Edward Elgar co-authored with Shunsuke Managi in 2009.

Shiro Hori he is a professor at Fukuoka University. He has worked as Japanese government member for 30 years. His research field includes international energy and environmental regime, climate change and energy security. Since 2011, he has acted as an expert for Japanese government on climate negotiation and energy-environment issues. His publication includes, "The Evolution of International Environmental Regimes", *The Waseda Journal of Social Science*, (2015), "The role of CSR in promoting companies energy saving actions in two Asian cities", *Energy Policy*, (2014).

Ryoji Hasegawa is an Assistant Professor at Osaka International University. He has Ph.D. in Economics from Kobe University. He majors in environmental economics and regional economics. His work has been about energy and climate policy, waste management, and input-output analysis. He has work at Institute for Global Change Adaptation Science, Ibaraki University as a researcher and Research Institute for East Asia Environments, Kyushu University as an associate professor.

Hidemichi Fujii is an Associate professor at Nagasaki University. He has Ph.D. in Arts from Hiroshima University. His research work has been about corporate environmental management, environmental innovation and productivity analysis. He has worked at the Graduate School of Environmental Studies in Tohoku University as Research Fellow of the Japan Society for the Promotion, and FUJITSU LABORATORIES LTD. as a researcher.

Mami Shinozaki is an Associate Professor of Kyushu University. She specializes in energy saving policy management and environment education issues. Shinozaki is fluent in Chinese and has lived and worked in China for 10 years. She has experienced at the Japan Airlines, Hitachi Construction Machinery, the Consulting Firm, Industry-University-Government collaboration

Management Center of Kyushu University as an international project manager and a researcher, and Research and Education Center of Carbon Resources of Kyushu University as a researcher. She is in charge of the curriculum of East Asia Environmental Strategist Training Program and the subject "Environmental Project Management", "Principles and Exercise of environmental pollution management" and "Internship for Environment in East Asia" at present.

Kenichi Tsukahara is a professor at Faculty of Engineering of Kyushu University. He has B.Eng from Kyushu University and Ph.D. from the University of Pennsylvania. He experienced senior positions in the government of Japan, Asian Development Bank, and Japan International Agency. He is now serving as a member of the Science Council of Japan and the chair of Water Related Disaster Risk Management Committee in the World Federation of Engineering Organization.

Takeru Sakai is a professor at Kyushu University. He has B.Eng., M.Eng., and Dr.Eng. from Kyushu University. His research work has been about Urban Design and Landscape Design, City Planning and Architecture. He has worked at the Sogo Kenchiku Architects, ACROS Fukuoka Project at Fukuoka Prefectural Government, Faculty of Engineering, Kyushu University as a research associate, and New Campus Planning Office as a professor. He has also worked at Department of Architecture and Graduate School of Engineering.

Masaki Yokota is an assistant professor at Kyushu University. He has B.Eng., M.Eng., and Doctor of Eng. from Kyushu University. He has worked for IDEA Consultants, Inc. His research work has been about coastal engineering issues.

Hiroki Nakamura is a Research Fellow at Department of Urban and Environmental Engineering, Kyushu University. He has B.Eng., M.Eng., and Ph.D. in Eng. from the Tokyo Institute of Technology. His research work has been about energy, environment, and transport policy and management issues. He has worked at the Energy and Environment Department, Japan Productivity Center, the 'Multidisciplinary Education and Research Center for Energy Science, Tokyo Institute of Technology Global Center of Excellence (GCOE) Program' as a research assistant, Research Institute for East Asia Environments, Kyushu University as an assistant professor, and a lecturer at the University of Kitakyushu.

Koichiro Aitani is a design expert with international practical experience in USA, Europe, Middle East and Asia, on complex residential, commercial, mixed-use and large-scale architecture and urban projects, such as NATO headquarter, Cathedral of Christ the Light with Skidmore, Owings and Merrill, LLP of San Francisco and London. He is a certified professional (1st Class Architect, Japan), also an Associate Professor at the Department of Architecture of Texas A&M University, and a Visiting Professor of Kyushu University. Having over 20 years of practical experience in Architecture and Urban Design with 7 years of teaching, He brings an incredible collaborative spirit and breath of scientific knowledge to every project he touches, with sustainable aspect. Research Interest: Architecture and Urban Design, High-rise Building, Urban Regeneration, Architecture Education.

Part I
Environmental Knowledge

Chapter 1
Asia's Local Air Pollution and Impacts on Global Climate Change

Nobuhiro Horii

Abstract Asia has become one of the global economic centers responsible for nearly 30 % of global GDP as of 2011. This may not be too surprising when we consider the fact that 55 % of global population is concentrated in this area. However, considering the fact that Asia accounted for only 15 % of the global economy in 1973, it can be said that the change over the past 40 years has been quite considerable. At the same time, energy consumption increased along with economic growth. Furthermore, now Asia has a huge share of air pollution and greenhouse gases emission. The purpose of this chapter is to organize the data of the current situation where energy consumption is increasing along with economic growth in Asia and to have an overview of local air pollution and global environmental problems brought about as a result of energy consumption. We will analyze the energy structure in Asia which has low energy efficiency and large environmental impacts. However, as for local air pollution, the indications are showing that it has been on a course of improvement in recent years. Finally, the future outlook is examined based on that given by the International Energy Agency (IEA).

Keywords Air pollution • Climate change • Energy and development • Asia

1.1 Economic Growth and Energy Consumption in Asia: What Is the Key to Energy Saving?

First of all, we need to confirm the position of Asia in the whole world economy, energy, and environment. As shown in Table 1.1, the GDP of Asia accounted for 27.4 % of the global economy as of 2011, representing a huge increase in the impact on the global economy compared with the 15.1 % figure in 1973. Big changes can be seen among the shares of each country. Japan produced more than 70 % of Asian GDP in 1973; however, its share decreased to 30.4 % in 2011, indicating multi-polarization, in particular the robust growth rates of China and India. The GDP of

N. Horii (✉)
Faculty of Economics, Kyushu University, Fukuoka, Japan
e-mail: horii@econ.kyushu-u.ac.jp

T. Shimaoka et al. (eds.), *Basic Studies in Environmental Knowledge, Technology, Evaluation, and Strategy*, DOI 10.1007/978-4-431-55819-4_1

Table 1.1 Economic, energy, and CO_2 emission indicators for the world, regional, and each country in Asia

	1973			2011		
	Real GDP (billion US$)	Energy consumption (million toe)	CO_2 emission (million tons)	Real GDP (billion US$)	Energy consumption (million toe)	CO_2 emission (million tons)
North America	5,995	1,887	5,070	16,296	2,421	5,726
Latin America	1,615	275	545	5,282	809	1,606
EU	9,305	2,316	6,732	20,537	2,900	6,433
EU (OECD)	8,040	1,371	3,993	17,965	1,719	3,650
Africa	494	212	273	1,760	700	1,048
Middle East	553	58	152	2,378	671	1,635
Oceania	438	65	190	1,356	136	397
Asia	3,260	1,109	2,427	17,952	5,036	13,841
Japan	2,293	321	891	5,464	457	1,174
China	220	427	975	6,503	2,723	8,561
Hong Kong	31	3	10	240	14	48
Taiwan	45	12	39	466	110	262
Korea	102	21	70	1,052	260	577
Singapore	19	4	10	244	35	62
Brunei	6	0	1	13	3	9
Indonesia	90	38	33	755	194	399
Malaysia	27	6	14	259	78	189
Philippines	55	17	27	207	31	78
Thailand	39	15	22	319	121	229
Vietnam	11	15	18	113	62	132
India	217	164	207	1,813	759	1,801
World	21,659	6,105	15,962	65,560	13,031	31,811

Source: The Energy Data and Modelling Center, IEEJ (2014)

China accounted for 36.2 % of the whole of Asia, surpassing Japan in 2011. The GDP of India in the same year was 10.0 %.

China is not only one example, but a lot of countries in Asia are also on a course of economic development based on industrialization, which mandates a rapid increase in energy consumption. China, which accounts for the largest energy consumption in the area, surpassed the USA in 2010 to become a nation with the highest energy consumption in the world. However, the increase in energy consumption is more than six times as much, while the size of its economy increased by nearly 30 times. On the other hand, India has increased its economy by over eight times, while its energy consumption grew by nine times. Similarly, the rate of energy consumption has increased more than the rate of economic growth in Korea, Indonesia, Malaysia, and Thailand. On the other hand, Taiwan and Japan have improved their energy efficiency.

When GDP-specific energy consumption (energy consumed to produce 100 million dollars worth of GDP) is focused, China showed a huge improvement in energy efficiency, from 194,000 tons in 1973 to 42,000 tons in 2011. However, compared with the Japanese GDP specific energy consumption of 8,000 tons in 2011, it still remains at lower than one fifth of the level. When compared at the level as of 1973, the GDP-specific energy consumption of India was 76,000 tons, at a better standard than China was. Though India has achieved energy efficiency improvement, its GDP specific energy consumption in 2011 stayed at 42,000 tons, at the same level as China was.

Energy efficiency improvement and energy saving that China has accomplished in less than 40 years deserve a high valuation, though their energy efficiency level still remains low. In the background, it is pointed out that China was under a planned economic system prior to 1978 when economic reform was launched. Under a planned economic system, users just received distributed energy according to government plan without paying any cost. Therefore users used up all the distributed energy because of no incentive for energy saving, resulting in serious waste of energy. However, the market mechanism was introduced into energy trading after the 1980s. Furthermore, the price level has been raised gradually. So now, the price of coal, which is the main energy source in China, is determined in reflection of the market balance between demand and supply (Horii 2014a, b). As a result, users became aware of saving energy costs, which led to increasing energy efficiency as a whole.

Meanwhile, India began to grow out of its planned economy in the 1990s, somewhat later than China, but the price of energy is still kept low in consideration for the low income class (or the demand of industries hiding behind it) (Horii 2013). A massive blackout involving 600 million people occurred in July 2012 because of a chronic deficiency in investment for electricity generation facility reinforcement due to the fact that the price of electricity is kept artificially low. Furthermore, the price of coal, the fuel for power generation, is kept artificially low by government policies, which leads to deficient supply. Importing coal from overseas to make up for the deficiency is not economically viable due to high imported coal prices and the low price of electricity. As we have seen here, a low energy price can be a restricting factor on energy supply, and in addition to that, demand will not decrease because of artificial suppression of price rises during deficient supply This results in severe shortage of supply and increase of energy waste. The difference between China and India indicates the importance of market function to promote energy saving.[1]

[1] Energy theft is said to amount to nearly 20 % of power generation in India despite the low price of electricity. Therefore, some argue that raising the price of electricity is difficult in practice. However, it is better to assist low-income people with welfare policies such as cash handouts to match the amount of electricity price increase. By doing so, low income people also will have an incentive for electricity saving so that they can keep more of the handout at hand (and the unused money may be spent on children's education, etc., which is better for society). It is expected that

Table 1.2 Subsidy for lower energy price in Asia (2011)

	Ratio to GDP				Ratio to government's revenue			
	Oil	Electricity	Gas	Coal	Oil	Electricity	Gas	Coal
China	0.00	0.15	n.a.	n.a.	0.00	0.68	n.a.	n.a.
India	1.25	0.32	0.17	0.00	6.75	1.72	0.90	0.00
Indonesia	2.58	0.66	0.00	0.00	14.51	3.69	0.00	0.00
Thailand	0.15	1.64	0.14	0.25	0.66	7.24	0.61	1.08
Malaysia	1.24	0.33	0.31	0.00	5.67	1.49	1.41	0.00
Philippines	0.00	0.00	0.00	0.00	0.00	0.00	0.00	0.00
Myanmar	0.54	n.a.	n.a.	n.a.	9.35	n.a.	n.a.	n.a.
Laos	0.00	n.a.	n.a.	n.a.	0.00	n.a.	n.a.	n.a.
Cambodia	0.00	n.a.	n.a.	n.a.	0.00	n.a.	n.a.	n.a.
World	0.30	0.22	0.16	0.01	0.91	0.64	0.48	0.03

Source: IMF (2013)

Further still, typical cases where energy saving was compromised due to suppressed energy prices are seen in Indonesia and Malaysia. Both were once prominent oil-/gas-producing nations. They maintained a system to keep domestic energy prices low based on the idea of returning the benefit from natural resources to their citizens. However, the growth in energy production has been stagnating, and the subsidy, along with rising international prices, imposes a heavy burden on the economy and finance. As is seen from Table 1.2, a lot of countries pay an energy subsidy in Asia, which is one of the main factors holding back improvements in energy efficiency.

1.2 Increasing Impact on Climate Change: High Dependency on Coal

Next, the impact on the environment is examined. As indicated in Table 1.1, the Asian share in world CO_2 emission shows a large increase, from 15.2 % in 1973 to 43.5 % in 2011. Unlike traditional air pollutants, there is no abatement technology which is economically available to reduce CO_2 emissions completely.[2] Although CO_2 is a major component of greenhouse gas, it is inevitably emitted into the air accompanied by use of fossil fuels. Therefore there is a high correlation between CO_2 emission and energy consumption. Yet in those countries where alternative

energy wasting can be eliminated as a whole, leading to a good result where supply deficiency is eased.

[2] Even with carbon capture and storage (CCS) technology which is considered most promising, it cost 4,200 yen per ton CO_2 in Japan as of 2013. The target is to decrease the cost to 1,000 yen per ton to achieve commercialization in the 2020s.

energy use is developed, such as nuclear energy and renewable energy, the amount of emission is lower. Moreover, in those countries where fossil fuel with a higher carbon content, i.e., coal, is mainly used, the emission is greater than those countries where gas is often used.

As is mentioned previously, Asia's share of CO_2 emission in the world (43.5 %) is far greater than its GDP share (27.4 %) mainly due to low energy efficiency. In addition, the CO_2 emission share is slightly larger than the energy share (38.6 %) because carbon intensity in the energy structure is high, especially when the dependency rate on coal as an energy source is high. As shown in Table 1.3, 67.2 % of world coal consumption as of 2011 was in Asia. Considering the fact that the proportion in 1973 was a mere 22.0 %, this increase indicates coal was the fuel of choice when trying to meet the increasing demand for energy as Asia was about to launch its economic takeoff. When the breakdown in each country is focused, coal consumption in China is predominantly excessive. Coal consumption in China is as much as 73.3 % of the Asian total and 49.2 % of the world's total. Consumption in India is also very high. Coal consumption in Taiwan, Korea, and Indonesia is also fairly high, though the quantity is much lower compared with these two countries.

What we should focus on is the fact that consumption of coal and oil in Europe is decreasing, indicating a shift to natural gas and nuclear energy. In North America, the rate of conversion from coal to gas is expected to make a huge leap due to the expansion of shale gas production. On the other hand, a lot of countries are expected to increase their dependency on coal, which will be examined later according to the future prediction of the IEA. In particular, a large increase is predicted in India. Furthermore, coal is expected to become the main energy source in Indonesia due to restrictions placed on domestic oil/gas production.

Meanwhile, coal consumption volume in China will continue to increase for a while, but the ratio of coal in primary energy is expected to decrease because the growth of other energy sources will be correspondingly greater. It is also partly because coal price had been on rise until 2012 since around the mid of 2000s, due to coal pricing reform, and coal is now not very specially cheap energy, compared with other energies, as before (Horii 2014a). In addition, China is also actively introducing renewable energy and now is the largest country with the largest wind power capacity in the world. As is seen in Table 1.3, China holds the highest level of other energy sources including renewable energy.[3] Hydropower and nuclear capacity are also expanding rapidly.

[3] By the way, this figure is high also in Africa and India. This is because consumption of so-called traditional biomass energy, that is, firewood and animal manure, makes up a large proportion. On the other hand, the figure in China is mostly modern renewable energy, such as wind power and solar energy.

Table 1.3 Energy consumption structure in the world, regional, and each country in Asia (unit: million toe)

	1973						2011					
	Coal	Oil	Gas	Nuclear	Hydro	Other	Coal	Oil	Gas	Nuclear	Hydro	Other
North America	326	897	552	27	40	45	499	868	652	238	60	104
Latin America	8	154	33	–	9	71	38	372	189	8	66	136
EU	773	1,068	357	23	42	53	537	824	1,006	313	68	152
EU (OECD)	425	732	135	19	29	31	304	579	428	236	43	129
Africa	38	41	3	–	3	127	107	148	94	4	10	337
Middle East	1	43	14	–	0	0	11	319	338	0	2	1
Oceania	24	31	4	–	2	4	50	47	30	–	4	5
Asia	330	400	16	3	14	346	2,536	1,198	478	110	91	623
Japan	58	249	5	3	6	–	107	206	100	27	7	10
China	205	52	5	–	3	162	1,859	442	108	23	60	231
Hong Kong	0	3	–	–	–	0	8	4	2	–	–	0
Taiwan	2	9	1	–	0	–	41	40	15	11	0	3
Korea	8	13	–	–	0	–	80	94	42	40	0	4
Singapore	0	4	–	–	–	0	–	25	8	–	–	2
Brunei	–	0	0	–	–	0	–	0	3	–	–	–
Indonesia	0	11	0	–	0	27	32	73	35	–	1	53
Malaysia	0	4	0	–	0	2	16	28	29	–	1	4
Philippines	0	9	–	–	0	8	8	12	3	–	1	7
Thailand	0	7	–	–	0	8	18	47	31	–	1	24
Vietnam	2	4	–	–	0	9	16	21	7	–	3	15
India	36	24	1	1	2	100	326	166	50	9	11	197
World	1,501	2,816	979	53	110	646	3,776	4,136	2,787	674	300	1,358

Source: Same with Table 1.1

1.3 Current Situation of Traditional Air Pollution: Basically Improving

Based on the analysis on the feature of the energy structure in Asia, let us confirm the current emission situation of local air pollutants. Table 1.4 shows the SO_2 concentration in major cities of Asia. 2010 data are not available for some cities; however, looking at comparable cities, an improvement can be observed in all cities except for Jakarta (Indonesia) which showed worsened SO_2 concentration. Although the most serious pollution is taking place in China, compared with Tokyo, Beijing and Shanghai, more than five times higher level of pollution is observed, but these cities showed an improvement in SO_2 concentration when compared with the figures in 2010. Beijing had succeeded in lowering to less than half of the concentration. This is because major SO_2 pollution comes from fixed sources such as power generation plants and factories. In China, such fixed pollutant sources are often located in the northern cities such as Beijing, but the introduction of flue-gas desulfurization (FGD) equipment into power generation plants was promoted during the 11th Five-Year Plan period (2006–2010). The introduction of FGD equipment reflects a major reduction in SO_2 emission. Both Beijing and Shanghai are still at a higher standard, but already the situation is more serious in Jakarta. The worsening situation in Jakarta can be considered to be related to the development of industrialization in Indonesia in recent years, in particular the increase in the number of coal-fired power plants.

Table 1.4 SO_2 concentration in major cities of Asia (unit: annual average ppm)

Country	City	2000	2010
Japan	Tokyo	0.004	0.002
	Osaka	0.006	0.004
Korea	Seoul	0.006	0.005
China	Beijing	0.025	0.011
	Shanghai	0.016	0.01
	Hong Kong	0.01	0.004
Taiwan	Taipei	0.004	0.004
Philippines	Manila	0.010 (2001)	n.a.
Vietnam	Hanoi	0.003	n.a.
	Ho Chi Minh	0.006 (2003)	n.a.
Malaysia	Kuala Lumpur	0.004	n.a.
Indonesia	Jakarta	0.004	0.018
Singapore	Singapore	0.008 (2001)	0.004
Thailand	Bangkok	0.004 (2001)	0.002
Myanmar	Yangon		n.a.
India	Delhi	0.006	n.a.
	Mumbai	0.003	n.a.
	Kolkata	0.005	n.a.
	Chennai	0.003	n.a.

Source: Various sources

Table 1.5 NO_2 concentration in major cities of Asia (unit: annual average ppm)

Country	City	2000	2010
Japan	Tokyo	0.028	0.02
	Osaka	0.024	0.017
Korea	Seoul	0.035	0.034
China	Beijing	0.035	0.028
	Shanghai	0.044	0.024
	Hong Kong	0.046	0.059
Taiwan	Taipei	0.031	0.029
Philippines	Manila	0.022 (2001)	n.a.
Vietnam	Hanoi	0.005	n.a.
	Ho Chi Minh	0.043	n.a.
Malaysia	Kuala Lumpur	0.011	n.a.
Indonesia	Jakarta	0.015	0.011
Singapore	Singapore	0.016 (2001)	0.012
Thailand	Bangkok	0.031 (2001)	0.02
Myanmar	Yangon		n.a.
India	Delhi	0.015	n.a.
	Mumbai	0.015	n.a.
	Kolkata	0.040 (2002)	n.a.
	Chennai	0.008	n.a.

Source: Various sources

Next, let us look at the situation of NO_2 concentration by referring to Table 1.5. Cities with available data showed improvement also in NO_2 concentration from 2000 to 2010 with the exception of Hong Kong Central. The number of NO_2-generating fixed sources such as power generation plants and factories increased during this period. And unlike SO_2, emission from mobile sources such as automobiles accounts for a large proportion. The number of privately owned vehicles in China underwent a rapid increase, from 16.09 million in 2000 to 78.02 million in 2010 as shown in Table 1.6. In the whole of Asia, the number of vehicles rapidly increased from 139.78 million in 2000 to 244.91 million in 2010, accounting for 23.3 % of the world's total (18.2 % as of 2000). China's contribution rate to the increase in the whole of Asia is a dominating 60.8 %, but other countries including Indonesia, Malaysia, Thailand, and India show high growth rates as well. Though the number of privately owned vehicles is increasing, NO_2 concentration is decreasing. The reason for this is considered to be that older vehicles with higher discharge rates were largely replaced by newly disseminated vehicles with advanced NO_2 abatement technology during this period.

The NO_2 emission level in Beijing and Shanghai is not at the worst level in Asia. The figures are worse in Hong Kong, Taipei, and Seoul. The reason for high NO_2 concentration in these cities is largely because of the impact of vehicle exhaust emission, which is usually a major cause of NO_2 concentration in major cities. China has a great deal of pollution from fixed sources; however, power generation plants and factories are often located in the suburbs, geographically distant from the

Table 1.6 Number of vehicles owned in the world, regional, and each country in Asia (unit: thousands)

	2000	2010
North America	239,046	263,114
Latin America	54,681	94,355
EU	284,660	372,718
EU (OECD)	238,065	290,079
Africa	19,646	29,117
Middle East	14,168	28,651
Oceania	14,950	18,701
Asia	139,775	244,906
Japan	72,370	74,997
China	16,089	78,018
Hong Kong	490	781
Taiwan	5,549	6,826
Korea	12,060	17,941
Singapore	561	798
Brunei	181	253
Indonesia	5,412	15,829
Malaysia	5,242	10,050
Philippines	2,434	3,119
Thailand	6,120	10,700
Vietnam	372	1,190
India	9,420	19,282
World	766,926	1,051,562

Source: Same with Table 1.1

center of the city. Therefore, their contribution to NO_2 concentration in cities is limited. Conversely, vehicle emission is discharged in the cities, having a direct impact on NO_2 concentration. These features, which are different from those of SO_2, are considered to be the reason for the high figure in Hong Kong and Taipei where traffic concentration is high. In Beijing, the size of the city itself is large enough to disperse traffic more or less. Moreover, in Shanghai, taking measures such as restrictions on issuing vehicle registration numbers to limit the number of vehicles entering the city may have an impact on NO_2 concentration.

The next item to consider is suspended particulate matter. As shown in Table 1.7, Beijing and Shanghai, both in China, have by far the most serious pollution among all Asian cities. It is easy to understand the fact that repeated occurrence of serious smog in China, in which not even 20 m of visibility could be secured, attracted worldwide attention. PM2.5, which is thought to be identified as a causative agent of smog, was not included in the object for monitoring as of 2010; therefore, no data is available to understand the time series variation of PM2.5. It was not until after 2012 when PM2.5 is officially monitored in China. The suspended particulate matter which has been conventionally monitored in the traditional environmental regulation is PM10 which has much larger particles than PM2.5. In fact, a lot of countries in Asia mainly focus on PM10 in their environmental regulations (even in

Table 1.7 PM concentration in major cities of Asia (annual average mg/m^3)

Country	City	2000	2010
Japan (SPM)	Tokyo	0.039	0.021
	Osaka	0.034	0.023
Korea (PM10)	Seoul	0.06	0.049
China	Beijing (PM10)	0.162	0.121
	Shanghai (SPM)	0.156	0.079
	(PM10 since June 2000)	0.1 (2001)	n.a.
	Hong Kong (PM10)	0.066	0.059
	Hong Kong (PM2.5)	0.044	0.036
Taiwan	Taipei	0.056	0.055
Philippines	Manila (PM10)	0.075	
	Manila (PM2.5 since 2001)	0.045 (2001)	0.033 (2009)
Vietnam (PM10)	Hanoi	0.126	n.a.
	Ho Chi Minh	0.5 (2001)	n.a.
Malaysia	Kuala Lumpur (PM10)	0.045	n.a.
Indonesia	Jakarta (PM10)	0.052	0.049
Singapore	Singapore (PM10)	0.08 (2001)	0.026
	Singapore (PM2.5)	0.021 (2004)	0.017
Thailand	Bangkok (PM10)	0.083	0.061 (2009)
Myanmar	Yangon	n.a.	n.a.
India	Delhi (PM10)	0.12 (2001)	0.26
	Mumbai	0.109	n.a.
	Kolkata	0.125	n.a.
	Chennai	0.09	n.a.

Source: Various sources

Japan, it was after 2009 when PM2.5 was included in the objects for environmental regulation).

From Table 1.7, a decreasing trend of PM10 concentration is observed in many cities in Asia except for Delhi. Although the absolute amount in Beijing and Shanghai is high, however, we can say that they have shown considerable improvement in the past 10 years.[4] Dust and soot are the main components of PM10 pollutants, and introduction of dust and soot removal devices including electric dust collectors and bug filters was enforced in power generation plants and factories. As a result, dust and soot emission in China showed a significant decrease, down from 11.65 million tons in 2000 to 8.29 million tons in 2010. However, suspended particulate matter includes sand dust (not only the expansion of desertification but also increased construction work in cities) and aerosol which is the

[4] As was described previously, official monitoring data of PM2.5 did not exist until after 2012; however, observation figures by NASA satellite data revealed that Beijing and Shanghai accomplished over 5 % improvement in annual average concentration when periods 2001–2003 and 2008–2010 are compared (Zell et al. 2012).

second transformation of SO_2 and NO_2. Therefore multiple factors need to be examined. The technology for reducing SPM can be introduced at a far lower cost than the cost of flue-gas desulfurization equipment or denitrification equipment. Technical countermeasures are expected to make advances.

We have seen the overview of the current situation of local air pollution in major cities in Asia. Different from the general impression, it should be stressed again that the pollution situation is improving in many cities in Asia. It is because those countries can afford to introduce environmental abatement technology owing to economic development, typically seen in the dissemination of flue-gas desulfurization (FGD) equipment in China. Also lowering NO_2 concentration is considered to come from advancement of NO_2 emission reduction technology for vehicles. On the other hand, final products of companies are shifted to those with same quality but with lower environmental impacts, for example, by use of information technology, bringing the effect of decreased pollution emission. As Asia continues to experience economic growth, it is important to adequately reflect the cost brought about by wasted energy and environmental pollution in pricing and to develop effective and practical environmental regulation in order to encourage the introduction of the environmental technology.

1.4 Future Outlook for Energy Demand and Environmental Problems in Asia

Finally, we will examine the future outlook of air pollution in Asia and global warming problems based on the forecast by the International Energy Agency (IEA). Table 1.8 shows the estimate of energy demand for 2025 and 2035 in Japan, China, India, and ASEAN.[5] Obviously China will be consuming nearly 1.7 times more energy than the USA, and its world share will reach a dominant 23.8 % in 2025. However, the annual average growth rate is forecast to be 2.4 % for the years 2011–2025. Considering the fact that China was expanding energy consumption at annual average rate of 9.1 % in the 2000s, the projected rate of increase is greatly reduced in comparison. The increase rate will be only 0.7 % in the period of 2025–2035, with a slight decline in the world share to 23.4 %. Meanwhile, the absolute amount of energy consumption in Japan and the USA will decrease in 2035 in comparison with that for 2011. The absolute volume of energy consumption in China will still increase in 2035; however, it is expected that policies and regulations to convert the

[5] IEA shows energy demand outlook in three scenarios: Current Politics Scenario where current standard of energy saving and environmental regulation is extended, New Policies Scenario where policies and regulation with high probability of introduction are included based on current international debate, and 450 ppm Scenario where CO_2 concentration is restricted to 450 ppm (= rise in temperature is maintained within 2°). IEA itself seems to consider probability highest in its New Policies Scenario. Therefore, this paper will introduce the New Policies Scenario as the future outlook.

Table 1.8 Energy demand forecast of Asian major countries and USA (unit: million toe)

	2025							2035						
	Coal	Oil	Gas	Nuclear	Hydro	Other	Total	Coal	Oil	Gas	Nuclear	Hydro	Other	Total
Japan	107	156	102	50	8	36	458	98	131	103	45	9	57	443
China	2,166	667	331	195	113	314	3,786	2,135	726	442	248	122	387	4,060
India	499	273	94	32	20	227	1,146	681	380	143	53	32	251	1,539
ASEAN	192	274	168	4	13	154	804	279	313	208	8	18	177	1,004
USA	437	728	630	233	26	210	2,264	411	614	646	241	27	304	2,242
World	4,312	4,548	3,576	979	430	2,030	15,877	4,428	4,661	4,119	1,119	501	2,558	17,387

Source: IEA (2013), IEA and ERIA (2013)

economic structure to one with better energy efficiency will be introduced, in a similar way to the process in Japan and the USA.

Meanwhile, India will continue to greatly expand energy consumption. The annual average growth rate of energy consumption toward 2025 in India is 3.0 %, and the 2025–2035 forecast is also 3.0 %. That is to say, India's economic development system at present, with a large energy and environmental burden, is expected to continue. Steady economic growth is also expected in ASEAN countries. Energy consumption growth is forecasted to be 2.8 % until 2025, and the period of 2025–2035 is 2.2 %. As a result, India will share 8.9 %, and ASEAN will share 5.7 % of the world total in 2035.

Let us look at the change of energy sources composition. China is forecast to experience a major transition between 2011 and 2025 and also 2035. The absolute volume of coal consumption is decreasing, and therefore, component ratio of coal in the primary energy in 2025 and 2035 will account for 57.2 % and 52.6 %, respectively, representing a huge decrease from the 73.0 % of 2011 indicated in Table 1.3. Total energy consumption is increasing though coal consumption is slowing down, which means other energy sources are increasing. Comparing Tables 1.3 and 1.8, the increased volume of oil and gas consumption is considerable. The consumption of oil will be almost twice as much in the period from 2011 to 2035. It seems to be expected that motorization will continue to advance, and demand for transportation energy will increase. In addition, gas is an alternative for coal, and currently fuel switch from coal to gas is positioned as a key measure against air pollution (especially against PM2.5). Therefore gas consumption is expected to grow more than four times as much as it is now. Nuclear energy is even higher, expected to grow more than ten times bigger. Although the incident at Fukushima Daiichi Nuclear Power Plant had a temporary effect to slow down nuclear power plant construction in China, it is now being accelerated again after reinforcements in safety control systems after 2012 (Horii 2014b). Another target to raise the ratio of nonfossil energy to 15 % by 2020 is also promoting construction of nuclear energy, along with hydropower generation, which will grow to more than twice as much.

As we have seen here, China is obviously moving away from coal-dominant energy structure, which will be diversified by other energy sources filling up the gap of coal. Meanwhile, coal dependency will be relatively enhanced in India and ASEAN countries. In India, coal accounted for 43.0 % of primary energy in 2011. The ratio is expected to rise to 43.5 % in 2025 and 44.2 % in 2035. The same can be said for ASEAN countries where the 16.4 % of 2011 will rise to 23.9 % in 2025. Although it is expected to go down to 18.3 % in 2035, however, it will continue to stay at a higher level compared with that in 2011. Coal will keep its importance as energy source for India and ASEAN countries to support energy demand accompanying their development at lower cost.

Such a difference in the future energy demand among Asian countries is naturally reflected in CO_2 emission. As shown in Table 1.9, it is forecasted that CO_2 emission will be reduced in Japan and as the USA. On the other hand, China will continue to greatly increase CO_2 emission, predicted to account for 28.2 % in the whole world in 2025. However, after that toward 2035, CO_2 emission in China

Table 1.9 CO_2 emission forecast of Asian major countries and the USA (unit: million tons)

	2025	2035
Japan	1,036	940
China	10,056	10,238
India	2,780	3,882
ASEAN	1,785	2,284
USA	4,982	4,489
World	35,722	37,242

Source: Same with Table 1.8

is expected to stay at the same level contributed by advancement of growing out of coal. In 2035, China will become predominantly the country of largest emission with a huge difference from the world-second USA. However, the rate of increase is expected to slow down.

On the other hand, CO_2 emission in India and ASEAN will continue to expand reflecting rapid growth in energy demand. 72.5 % of the global increase in CO_2 emission is expected to come from the increase in India between 2025 and 2035. Its share in the world will be 7.8 % in 2025 and 10.4 % in 2035, markedly narrowing the difference from that of the USA. In addition, the CO_2 emission share of ASEAN countries, which accounted for merely 3.7 % in 2011, will steadily rise to 5.0 % in 2025 and 6.1 % in 2035.

1.5 Conclusion

Asia is attracting attention as a center of economic growth in the first half of the twenty-first century, and therefore, it will have a huge impact on the whole world regarding energy consumption and furthermore on the local air pollution and climate change problems accompanied by increasing energy consumption. As has been pointed out in this chapter, Asia is responsible for massive impacts on environment out of all proportion to its economic size, due to low energy efficiency and high dependency on coal. However, a trend for improvement is observed in local air pollution, which is considered to be owing to advancement in introducing environmental abatement technology along with economic growth. In that respect, it is sufficiently possible to achieve both economic growth and environmental solution, whose success depends on system design such as an energy market and environmental regulation.

China will maintain a large presence with its huge energy consumption and environmental impacts not only in Asia but also in the world. However, the current policy trend to promote energy saving moving away from coal dependency and taking environmental countermeasures will also continue. On the other hand, India will continue to reinforce its current coal-dependent structure, increasing its world share of energy consumption and CO_2 emission, although the absolute volume will not be as high as that of China.

Introducing energy saving and environmental abatement technology will become the key for India, and also ASEAN countries, to continue their growth while preventing any increase of the environmental impacts. The experience of China is indicative on this point. It is possible to make a smooth transition to a more efficient growth system if subsidies for lower energy price are abolished, introduction of effective environmental regulation is achieved, and then energy pricing is set to reflect environmental impacts. Chinese companies are showing "Green growth" in environmental technology such as flue-gas desulfurization equipment or renewable energy sources such as solar panels and wind power generation turbines, providing products and solutions at overwhelmingly lower cost than companies in developed countries (Horii 2010). The rise of Chinese companies as suppliers in energy saving and environmental solutions will eventually provide for India and ASEAN countries options to reduce the environmental impacts at lower cost. Currently Asia is a region of mounting issues of energy consumption and environmental burden in the world. It is expected that energy saving/environmental countermeasure technology will be sent out from Asia due to the fact that they are well placed in the field to deal with these issues.[6]

References

Horii N (2010) China, emerging as "new and big market for environmental industry" and international competitiveness of Japan in energy saving and environmental fields: strategy for Chinese market, key to green innovation of the future. Monthly China Economy, June ed. 2010. Japan External Trade Organization, Tokyo, pp 35–60 (In Japanese)

Horii N (2013) Energy supply. In: Uegaki A, Tabata S (authors and ed) Sustainable economic development in major nations in Eurasia region. Minerva Shobo Co. Ltd., Kyoto, pp 187–213. (In Japanese)

Horii N (2014a) Chapter 11: Energy: pricing reforms and the end of low energy price. In: Watanabe M (author and ed) The disintegration of production: firm strategy and industrial development in China. Edward Elgar, Northampton, pp 307–333

Horii N (2014b) China rising as a major nation of nuclear power – background and risk of rapid growth. In: Takahashi N (ed) "Nuclear" in Asia and Us – observing Fukushima. Keio Institute of East Asian Studies, Keio University Press Inc., Tokyo, pp 133–177. (In Japanese)

IMF (2013) Energy subsidies reform: lessons and implications. http://www.imf.org/external/np/pp/eng/2013/012813.pdf

International Energy Agency (IEA) (2013) World Energy Outlook 2013. OECD/IEA, Paris

[6] Needless to say, Japanese companies are expected to contribute to tackle with these issues; however, they are not quite successful in marketing due to the high cost as solution for Asia. It will be necessary for them to develop products with dramatically lower cost through collaboration with Chinese companies. Refer to Horii (2010) for details.

International Energy Agency (IEA) and Economic Research Institute of Asia (ERIA) (2013) Southeast Asia Energy Outlook 2013. OECD/IEA, Paris
The Energy Data and Modelling Center, IEEJ (2014) Handbook of energy & economic statistics in Japan. The Energy Conservation Center, Tokyo
Zell E, Weber S, de Sherbinin A (2012) Bottom up or top down? Another way to look at an air quality problem. In: News and insights, environmental performance index website. http://epi.yale.edu/the-metric/bottom-or-top-downanother-way-look-air-quality-problem

Chapter 2
Issues and Problems Regarding Water in Developing East Asia

Tetsuya Kusuda

Abstract In this chapter, the concepts of water circulation on the earth and scarcity of water resources are explained first. Then, water problems and issues are summarized, including the fact that polluted water does not function as a water resource. In addition, a number of water problems and issues (i.e., flooding, water scarcity, sea level rise, and water pollution) are systematized. Then, the impact of such issues on the degradation of water quality and difficulty in accessing potable water are treated. Finally, the deficiencies in existing legal systems, financial predicament, and poor governance are explored as trigger elements. Most Asian countries are subject to monsoon rainfall patterns, and many have tropical climates. These, of course, are completely different from the climates found in Europe and North America. Accordingly, Asians need to develop our own methods to cope with Asian climate, particularly with typhoons (cyclones) and periods of heavy rainfall. In conclusion, the notion of Integrated River Basin Management is explained to promote the wiser use of water.

Keywords Water resources • Flood • Drought • Water pollution • Water problem • Water issue • Asia • Population increase • Urbanization • Economic development • Climate change • Sanitation • Integrated river basin management

2.1 Introduction

Clean, freshwater is essential for human survival and activities and the maintenance of ecosystems on land. Fresh water, however, makes up less than 3 % of all the water on the globe. Moreover, only 0.01 % of all water is in lakes and rivers available for use by humans and other living things. For marine organisms, clean seawater is also essential.

Water is a circulating substance, meaning that, on the human scale of time, the supply can never be exhausted. Evaporation of a small portion of the seawater on our planet (total volume 1338×10^6 km^3) provides 87 % of total evaporation, which is the major source of fresh water. Water conveys substances through circulation,

T. Kusuda (✉)
Kyushu University, Fukuoka, Japan
e-mail: kusuda@kyudai.jp

T. Shimaoka et al. (eds.), *Basic Studies in Environmental Knowledge, Technology, Evaluation, and Strategy*, DOI 10.1007/978-4-431-55819-4_2

from the hydrosphere via the atmosphere to the geosphere. The hydraulic retention time of water in the atmosphere, slightly less than 10 days, suppresses atmospheric pollution. On the other hand, the retention time in the sea, about 3000 years, brings about pollution of the abyssal regions.

The circulation patterns of water sometimes bring about floods and, at other times, droughts. Human beings have struggled to cope with such extremes since time immemorial. Recent extreme weather conditions have obliged us to experience what we have not ever before experienced. Increasing human population and activities, rising living standards, poor governance, and enclosed water resources also lead to new experiences regarding water stress.

Water usage does not mean real consumption because it cycles, but pollution does. Polluted water does not function as a renewable resource. Polluted water can be made clean again; this is technically possible, but too costly on massive scales to be sustainable. Even in the future, development of more suitable technologies may not make it economically viable.

The importance and roles of water in societies are exemplified in the following visionary statements in history: "To control China, one must first control water" – Guan Zhong, a politician and statesman during the Spring and Autumn period of Chinese history (c 720–645 BC), and "He who solves water problems in the world deserves two Nobel Prizes, one for Peace and one for Science" – former President of the USA, the late John F. Kennedy (1917–1963). Water is necessarily a major and increasingly important concern in any movement toward a sustainable society.

Due to the UN MDGs (Millennium Development Goals 2000), which we are required to achieve by 2015, about 500 million people on earth lack access to clean drinking water. About 1.2 billion people lack basic sanitation services, that is, provisions for treatment of wastewater and solid wastes. More than 200 million people regularly suffer from flooding, and most of these people are concentrated in Asia and Africa. Even though the current MDGs have not yet been achieved, we are still in need of new goals for the more distant future.

2.2 Structure and Elements Regarding Water Issues and Problems

2.2.1 Current Water Issues and Problems

There are a number of issues and problems regarding water. Regarding quantity, we must be concerned with flooding and lack of water resources; for storage, we face challenges from rising sea level. With transport, access to water resources, transport of pollutants and sediments, and the spread of pathogens complicate our lives. A number of social problems (e.g., poor governance, poor sanitation, insufficient capacity building) complicate all other problems. We face pollution factors including acid rain, eutrophication, and oligotrophication (excessive and inadequate nutrients, respectively). All these, singly and in concert, result in major ecological problems, leading to the degradation of biodiversity. Direct influences on water problems may lead to secondary problems in chain reactions.

Most of the countries in the world, apart from those in Asia, have ever experienced precipitation levels of 2000 mm per year, nor a true monsoon climate. In worldwide congresses and conferences, the intrinsic problems of water in Asia are sometimes not reflected in the main themes. Countries in Asia have social problems regarding population and economics that are different from those on the European and American continents. Countries in the monsoon and tropical zones where the climate is different from European and American countries have gotten news out reporting their water issues and problems, especially flooding.

2.2.2 Structure of Water Issues and Problems

Problems regarding water emerge in those areas where people live and/or items exist upon which people are working. The places where people cannot get any water have no problems regarding water, because there are almost no living things. Numerous life forms may exist within a given environment with dynamic stability. Even if the local environment undergoes change in its water balance, it would not cause severe problems so long as it changes within the limits of the physical and social resilience of its inhabitants. Changes that go beyond these limits would be expected to cause problems. This means that problems regarding water do not result from the total quantity of existing water, but from changes in the quantity. The characteristics of issues and problems regarding water depend on climate, morphology, and history of social conditions.

There are various factors that could cause the impacts we fear from changes related to water. In the future, factors causing impacts will likely include:

Rapidly growing populations
Urbanization
Uncertainties about the impacts of global climate change and weather extremes
Unexpected change in atmospheric and surface water circulation
Possible conflicts regarding overuse of freshwater and natural resources in coastal zones
Enclosure of water resources, especially on transboundary rivers, such as construction of reservoirs immediately upstream of the border
Poor governance
Lack of investment funds for management, construction, rehabilitation, and research

Looking toward the future, we may be able to solve such problems with technology. New applications of technology may reduce water problems, perhaps by better construction of reservoirs and purification followed by reuse. Taken together, these may decrease water scarcity. However, the technological development is supported by investment arising from the benefit of economic development. It also seems that it is not easy to solve all problems in this way, because economic development itself may bring about pollution in its growth stage. Industrialization is deeply concerned with economic development and just as deeply impactful on the environment. An economic surplus produced by development should be invested to improve the environmental situation. Thus, degradation of the environment can be reduced by means of economic development, which itself is one of the causes of the degradation of the environment.

Past experience indicates that a time lag exists between causes and the emergence of resultant impacts. To avoid following in the steps of our predecessors (advanced countries), we need a new advanced paradigm to prevent pollution in the early stages of development. This will require a lower level of total investment to maintain a good environment, as future development proceeds. Taking late countermeasures to pollution may result in higher GDP, but taking preventive measures brings about a better society where people will not have to experience heavy pollution. We would like to construct a society where pollution is not a by-product of economic development.

2.2.3 Characteristics of Water Issues and Problems

Water issues and problems have some fundamental characteristics that may provide means to provide simple solutions to at least some of them.

1. Locality

   The fundamental area unit of a water issue or problem is a river basin or a watershed. The circumstances of any one river basin regarding climate, population, land use, and industry are always different from those of adjacent river basins. The fundamental area unit is enlarged, in cases where the water provided is conveyed a long distance, across the boundaries of multiple basins, as is the case in San Diego, USA; Beijing, China; and Fukuoka, Japan. Generally speaking, specific issues and problems regarding water are local, but they are regarded as a global problem since there are numerous similar incidences all over the world.
2. Water shortage

   Water shortage results from an imbalance between existing water resources and water consumption and is not a direct consequence of the absolute amount of water present. When assessing the risk of water shortage, usage of more than 40 % of precipitation and a water supply of less than 1700 m^3 per capita per year in a region indicate a high probability. As a general rule for water basins occupied by humans, a water consumption ratio of 70:20:10 (for irrigation, industry, and daily living, respectively) is typical. Reallocation of water resources (among the three major categories) is one solution regarding drinking water shortage problems.

3. Conflicts

   History is littered with examples of competition and disputes over shared freshwater resources. It seems inevitable that tensions over water resources will increase as more and more people compete for a fixed water supply. Improving standards of living increase the demand for freshwater, and future global climate changes will make supply more problematic and uncertain. About 240 rivers, or other sources of fresh water, are shared by two or more nations or regions. This geographical fact has led to a geopolitical reality of disputes over shared international rivers such as the Mekong in Southeastern Asia; the Amur (the Heilong) in Eastern Asia; the Indus, Ganges, and Brahmaputra in Southern Asia; the Nile, Jordan, and Euphrates in the Middle East; and the Colorado, Rio Grande, and Paraná in the Americas. Problems regarding water quantity and/or quality of international rivers crossing national boundaries are called "transboundary problems," or in other words, "upstream-downstream problems."

   While regional and international legal mechanisms can reduce water-related tensions, these mechanisms have never received the support or attention necessary to resolve the numerous conflicts over water. Not all water resources disputes lead to violent conflict, but they have occurred in certain regions of the world, including Southern Asia and the Middle East. Chemical accidents that contaminate transboundary rivers can cause similar problems. The Danube River, under initiatives by the EU, has a water commission that coordinates the activities of all the nations inside the basin.

4. Dependence regarding economics

   Economic development followed by environmental degradation is not an example to be followed. This lesson has already been learned: water pollution should not be a by-product of economic development. Economic development with preventive measures aimed at preventing environmental degradation in advance is the example to be followed. Simultaneous investment in both economic development and environmental protection is a better way to reduce the total amount of investment required.

2.3 Development of Water Issues and Problems in Asia

2.3.1 *Population*

Asia is the world's largest and most populous continent. It covers 8.7 % of the Earth's total surface area and includes 30 % of its land area. With approximately 4.3 billion people, it hosts 60 % of the world's current population (2010). In 1950, the world population was 2.5 billion people and that of Asia was 1.4 billion people. Asia has had a high growth rate in the modern era. During the twentieth century, the Asian population nearly quadrupled.

According to the United Nations, many East Asian countries have experienced a decline in their total fertility rates over the past 50 years. In the period from 1965 to 1970, the average family had five or more children. From 2005 to 2010, the average shifted to less than 2.1 children per family. Based on historical growth rates and national calculations, it is estimated that between 2000 and 2050, national populations are expected to increase in every country of East, Southeast, South, and Central Asia, apart from Japan and Kazakhstan. Populations will double or nearly double in Pakistan, Nepal, Bangladesh, Afghanistan, Cambodia, and Laos. Growth rates will also be particularly high in India, Indonesia, Iran, Malaysia, Mongolia, Myanmar, the Philippines, and Vietnam.

2.3.2 Urbanization

Mainly because of rural-to-urban migration, Asia is the fastest urbanizing region in the world. According to UN estimates, the urban population in Asia will have nearly doubled in 30 years. By 2030, more than half of Asia's population will live in cities: a projected 2.6 billion people. The urban population of Asia will be higher than the urban population of all the other regions of the world. In 2025, the proportion of urban population is projected to be 52 % in East Asia, 53 % in Southeast Asia, and 45 % in South and Central Asia. More than half of the national populations will be urban in Brunei, China, Indonesia, Iran, Japan, Kazakhstan, Malaysia, Mongolia, North Korea, Pakistan, the Philippines, Singapore, South Korea, and Turkmenistan. This urbanization of national populations is reflected in the growth of Asia's largest cities. In 1975, there were only five megacities (with a population of more than ten million) in the world, and only two of these were in Asia. In 2015, 15 of the world's 23 megacities will be in Asia. This rapid and largely unplanned expansion of urban areas has resulted in serious problems of air, soil, and water pollution. Despite rapid urbanization, Asia's rural populations are also projected to grow. Most of the region's rural areas are already densely populated.

2.3.3 Economy

Asia has the second largest nominal GDP of all continents, after Europe, but it is the largest when measured in purchasing power parity. As of 2013, the largest economies in Asia are China, Japan, India, South Korea, and Indonesia. It is forecast that India will overtake Japan in terms of nominal GDP by 2020. By 2027, China is expected to have the largest economy in the world. The rapid, largely unplanned expansion of urban areas has also resulted in high rates of unemployment and underemployment.

2.3.4 *Climate*

1. Climate changes

   The water-retention capacity of the atmosphere increases by about 5 % with every 1 °C rise in temperature. Increasing temperatures are expected to make wet regions wetter and dry regions drier. Anticipated changes in atmospheric circulation patterns will push storm tracks and subtropical dry zones toward the poles. Most regions, as a result of rising temperatures, are losing snow cover on the ground (small area of deposit and earlier melting). Global warming may increase extremes regarding climate, including torrential rain, heavy snow, and drought that could lead to desertification.
2. Monsoon zone

   South and East Asia consist of tropical and temperate zones, part of which is in the Indian Ocean monsoon zone. The Asian monsoon zone includes subsystems, such as the South Asian monsoon, which affects the Indian subcontinent and surrounding regions, and the East Asian monsoon, which affects southern China, Korea, and parts of Japan. Since the monsoon zone has both rainy and dry seasons, the annual availability of water depends on the capacity of reservoirs.

2.3.5 *Floods*

As mentioned above, climate change will mean that floods and flash floods increase in magnitude and frequency. Single typhoons in Southern Asia have killed hundreds of thousands of people in Bangladesh through flooding of low-lying lands. Recently a large number of floods have been observed, such as that in Australia (23 January 2011); in Thailand (22 August 2011); in Myanmar, Thailand, Cambodia, Laos, Vietnam, China, the Philippines, and Indonesia (22–25 September 2013); and in the Philippines (11 November 2013).

2.3.6 *Groundwater and Water Resources*

Precipitation strongly affects existing water resources for people, but not directly. Between precipitation and supply of water to people, an infrastructure exists to purify and convey it. Regions where fossil groundwater is the predominant source, like the Loess Plateau in China, will face water shortage as the aquifers are depleted. On the Loess Plateau, the water table has been receding by more than 60 cm/year and is currently at a depth of several tens of meters. Such depletion of fossil groundwater makes society in these areas unsustainable.

2.3.7 Water Supply and Water Scarcity

Water scarcity has three aspects: lack of water resources (physical scarcity), greater demand than supply (social scarcity), and inadequate supply due to a lack of investment (economic scarcity). Water scarcity may occur in areas where there is plenty of rainfall or freshwater. Allocation of water resources and quality of water are also of importance. In China, to compensate for water shortage, water from the Yangtze River is transported across the Yellow River to the capital area.

From 1990 to 2010, millions of South Asians gained access to improved drinking water sources. However, this still leaves some 170 million people not using drinking water from improved sources. These unreached masses, mostly in rural areas, continue to rely on unprotected surface sources and unimproved sources. More people would be using improved drinking water sources if not for the fact that, at any given time, up to a quarter of South Asia's public water supply systems are not operational. This is due to breakdowns, poor maintenance, declining water sources (now exacerbated by climate change), and generally aging infrastructure.

2.3.8 Wastewater Treatment and Sanitation

In South Asia, the pace of sanitation improvement has not kept up with population growth. In 2010, the region had about 1057 million people without improved sanitation, some 30 million more than in 1990. Even worse, 692 million people in South Asia have no toilets at all and defecate in the open. In India alone, some 625 million people practice open defecation. From 1990 to 2010, South Asians increased sanitation coverage from 22 to 38 % of the population. During this 20-year period, 369 million people gained access to sanitation.

In South Asia, urban populations are twice as likely to have use of improved sanitation facilities as rural ones (60 % versus 28 %). Even so, urban sanitation coverage in the region has stagnated in recent years. The number of urban dwellers in South Asia having use of improved sanitation facilities rose from 134 million in 1990 to 196 million in 2010. While the richest quintile has 92 % coverage, the poorest quintile has only 4 % coverage.

For better sanitation and recycling resources, a toilet system with separate collection of urine and feces and a bio-toilet system for composting of raw sewage have both been commercialized.

2.3.9 Water Pollution and Sediment Transport

In Asia, rivers large and small are polluted with household wastes including pathogens, chemicals (e.g., detergents, pesticides, fertilizers), industrial waste, and oil. Lakes and coastal zones in Asia are often heavily polluted with a cocktail of harmful substances (dioxins, heavy metals, chemicals, fertilizers, industrial waste). The existing concentration of contaminants in the food chain brings about some health risks. In Japan, some semi-closed bays are so oligotrophic (poor in nutrients) that fisheries there are impacted. Lake Taihu in China exhibits pulses of extraordinary growth of phytoplankton, that is, algal blooms.

Rivers convey not only pollutants, but also sediments. Sediment transport in rivers is a significant part of an ecosystem, but natural (or accelerated by human activities) sediment deposition within rivers fills the drainage basins, thus increasing the risk of floods (e.g., Yellow River in China). On the other hand, sediment deposition in coastal areas provides essential littoral nourishment. In Japan, decreased sediment transport to the coastal zone causes environmental and ecological problems.

2.3.10 Acid Rain

Asian cities are among the most polluted in the world. Of the 15 largest cities on the planet with the worst air pollution, 12 are in Asia. Acid rain lowers the pH of surface water and has impacts on the biogeochemistry of ecosystems.

2.4 Integrated River Basin Management to Solve Problems and to Discuss Issues

Integrated river basin management (IRBM) has proven an effective management approach for solving the kinds of problems mentioned above and for providing forums for discussion of critical or significant water issues.

2.4.1 Elements of Integrated River Basin Management

The aim of IRBM is to provide effective responses to increasing demand for societies to develop economic (particularly industrial) activities while maintaining quality of life, accustomed activities, and aquatic environment in a desirable condition, at an achievable cost.

Effective basin planning requires as its foundation the intelligent use of information on the planning elements of population, land use, infrastructure, industry, and water circulation. These elements also include available technologies for water supply, wastewater treatment, disaster prevention, and protection of aquatic organisms. Also important are essential support, such as a legal system to control water quality and quantity; relevant technologies (GIS, modeling, measurement) for planning and estimation; involvement of stakeholders (agencies, businesses, and citizenry) and experts for decision-making; and capacity building for sustainability. Control of the amount of pollutants discharged from sources and water utilization considering water quality are also relevant elements.

2.4.2 Concepts for Management

To obtain solutions, it is necessary to integrate knowledge, effective action by organizations, and relevant legal framework (clear authority and enforcement). Seeking an optimum solution would seem the obvious target for management, but each element constituting the IRBM system remains dynamic, including the sense of value of the stakeholders, during any particular process. This means that "optimum solutions" are also dynamic or not stable. Usually we look for the least-complaint resolutions, which are also dynamic. There are numerous trade-off issues regarding water management, such as the human uses versus protection of biodiversity, allocation of water between categories (irrigation, industry, and daily life) and regions (upstream-downstream problems), cost allocation between present and future generations, and the higher income of inhabitants versus investment for construction of water pollution control facilities (often via taxation).

At present, the typical procedure by which water problems are solved occurs in the order drivers (causes), pressure on environment (pollutants), state change (pollution, ecological damage), impact to humans and ecosystem (consequences), and then response policy and countermeasures. A more modern and effective procedure would be drivers (causes), preventive measures (purification, effective allocation), and no state change. Concisely, the shift from DPSIR to DPN should lead to more efficient development.

2.4.3 Objectives of IRBM

The optimum solution for IRBM could be either a single one or a combination solution with weighted objectives:

Maximizing social benefit
Maximizing social fairness
Minimizing social loss

Minimizing social conflict
Maximizing biodiversity

In practice, effective solutions will likely be a combination of some of these objectives.

2.4.4 *Management Procedure*

To understand how water circulates in a basin system requires a broad range of expert knowledge (e.g., hydrology, hydraulics, sanitary engineering, urban planning, and wastewater treatment). To provide a context for the solution of water problems, a great deal of more information about conditions in the basin must be integrated (e.g., ecology, sociology, and economy). Finally, access to a wide range of technologies is required for effective IRBM.

A reasonable management procedure might involve the following objectives:

- To understand the natural water balance and artificial usage in the targeted watershed
- To monitor the environment long term and to provide the data needed to understand the current situation and future change
- To understand environmental conditions using modeling and analyses
- To understand social needs in the area
- To determine effective policy for management of the watershed
- To establish indices and criteria for decision-making
- To provide a future perspective on management using predictive models
- To acquire and/or modify the technologies necessary for management
- To explore management performance using simulation
- To build a partnership of stockholders within the basin and with relevant people and agencies outside the basin
- To arrange funding and/or investment to enable effective management

2.4.5 *Calculation Methods*

The following are the fundamental equations involved in IRBM. The objective function for maximizing social benefit is:

$$\text{Max } V = aV_{\text{pro}} + bV_{\text{ben}} - cV_{\text{risk}} + dV_{\text{ecol}} \tag{2.1}$$

where V is value; a, b, c, and d are coefficients of weight; *pro* is production; *ben* is benefit; *risk* is risk; and *ecol* is ecology, and the following assumptions apply (i.e., given that):

$$Q_{\text{tot}} = Q_{\text{agr}} + Q_{\text{ind}} + Q_{\text{urb}} + Q_{\text{ecol}} \tag{2.2}$$

$$V_{\text{pro}} = f_{\text{pro.agr}}(Q_{\text{agr}}) + f_{\text{pro.ind}}(Q_{\text{ind}}) \tag{2.3}$$

$$V_{\text{ben}} = f_{\text{ben.urb}}(Q_{\text{urb}}) \tag{2.4}$$

$$V_{\text{risk}} = f_{\text{risk}}(Q - Q_c) \tag{2.5}$$

$$V_{\text{ecol}} = f_{\text{ecol}}(Q_{\text{ecol}}) \tag{2.6}$$

where *tot*, *agr*, *ind*, *urb*, and *ecol* are total consumption, agriculture, industry, urban use, and ecology, respectively; Q is the water volume of net use; and f is the function.

In these equations, water quality is not considered. If necessary, the treatment cost to achieve a target quality could be added. Water draining from upstream farmland could be used again in downstream areas. The amount of water used in the whole basin is not equal to the summation of an amount in each plot. These equations are to be combined with the equations of water balance.

In the case of maximizing social fairness and biodiversity, V is read as fairness or biodiversity instead of value. In minimizing social loss and social conflict, Min V is used instead of Max V. In such calculations, each layer, or presentation for each element, is assigned a weight.

2.4.6 *Simplified Methods on IRBM*

When IRBM is too complicated to handle, or when the IRBM objectives are not integrated, the management method could be simplified by assigning at least some of the objectives to more specific, lesser categories (e.g., water resources management, water quality management, or coastal zone management).

2.4.7 *Example: Case Study of the Yellow River*

An outline is provided below; details are available elsewhere (Kusuda 2009).

1. Physical and hydrological characteristics of the Yellow River

   The Yellow River basin originates on the Tibetan Plateau, meanders through the northern semiarid region, crosses the Loess Plateau, passes through the Eastern Plain, and finally flows into the Bohai Gulf. The main stream of the Yellow River flows about 5500 km and drains water from an area of 753,000 km^2, of which farmland is 40 % (126,000 km^2). Moreover, irrigated farmland is 40 % of the total farmland, and the crop yield there is 40 Mton/year. The population density in this basin is 152 persons/km^2 and the average annual precipitation is 433 mm/year. The existing water resources are 49.1 billion m^3/

year and the river discharge was 36.3 billion m^3/year in 2008. The groundwater extraction was 12.8 billion m^3/year (2008), and sediment transport from the basin to the Bohai Gulf was 46 Mton/year (in 2008, but 1.16 billion ton/year from 1956 to 2000).

2. Data correction or adjustment

   The correction/adjustment factors used in relevant computations included land use, morphology (altitude, slope, slope length, soil characteristics, soil thickness, and vegetation), present and future distribution of the human population (based on National Plan), industrial products and the number of factories, climate data (precipitation, temperature, wind velocity, duration of sun light, evapotranspiration, and parameters on moisture content), water consumption (extraction, irrigation, and daily use), water table level, river flow rates, and water quality in rivers.

3. Conditions for calculation

   The targeted region was from the source to Huayuangou. The pixel dimension was 10×10 km. The number of grids was $182 \times 263 = 47{,}866$. The computation unit ($\Delta t$) was 1 h.

4. Basic equations

   For computation of the flow rate, the kinematic wave method and Manning equation were used; for soil moisture, the Richards equation and van Genuchten equation; and for groundwater, the Darcy equation.

5. Computation results

   Half of the observed data were compared with computation results for calibration, and the rest were used for validation (e.g., for such as the flow rate in rivers, the moisture content of soil, and the ground water level in each pixel). After validation, estimation was conducted by changing operational variables such as forestation, amount of irrigated water, irrigated area, irrigation efficiency, wastewater treatment ratio, water reuse ratio, and self-sufficiency of food. The operational variables were changed stepwise, except for self-sufficiency of food. The constraints used were stable water resources, arable land area, limited job-transfer ratio of farmers, labor share, and available water amount for agriculture, based on water quality. The five estimation areas were the upstream region, the middle stream region, the Wei River region, the Fen River region, and the downstream region.

6. Obtained results

   Based on the above conditions, crop yield, industrial products, costs of procurement, construction of infrastructure, and wage were estimated. Then, as targets, maximum benefit, wage increase, increase in employment, protection of ecosystem, and reduction of sediment were estimated via projections to 2030 and 2050.

Progress in forestation, with consequent reduction of sediment loss, led to a decrease in arable land. This decrease could be compensated for by an increase in irrigated farmland and irrigation efficiency and by job transfer of less effective farmers from agriculture to industry. In addition, increased allocation of water to farmland allowed farmers to expect greater crop yield. However, in this scenario, effective mobility of farmers and preparation of farmland for them (presumably in new locations) would be required.

References

Kusuda T (ed) (2009) Yellow river—water and life. World Scientific, Singapore, 175pp
Millennium Development Goals (2000) Home page of United Nations: http://www.un.org/millenniumgoals/Nov.2015

Chapter 3
Environmental Problems in China

Reiitsu Kojima

Abstract The peculiarity of China is described first, because this is strongly reflected in the deepening of environmental awareness and policy formation. The second point is the formative period of environmental policies, from 1973 to the mid-1990s. The third part is the change in 20 years after that. Finally, the fourth point is an overall assessment of environmental protection activities spanning 40 years. Please note that the word "environment" in this paper can indicate both "environment" in a narrow sense, such as air, water, and solid waste, and also broader issues of the ecosystem.

Keywords Negative historical heritage • Compressed environmental destruction • Weakness of economic methods in antipollution control policy • Authority power weakness of SEPA • Lack of mass antipollution resistance • Pollution expansion outside

3.1 Points Which Merit Extra Attention When Examining Environmental Ecosystem Problems in China

3.1.1 Negative Historical Heritage

The Qin and Han dynasties built a dazzling civilization more or less equivalent to that of the Roman Empire in the west, which lasted over 400 years. After that, the country underwent a period of almost continuous warfare, yet it still managed to maintain a high level of civilization. This was possible in those days due to an agricultural system with quite high productivity and a robust bureaucracy which accumulated wealth from agriculture. This agriculture enabled the country to support a huge population. However, the green forest in the mountains, rivers, thickets, and valleys supporting the residence of farmers was gradually cleared away in order to create grazing pastures for goats and sheep and to gain fuel and materials for agriculture and houses. Such action became a major factor of nature destruction.

R. Kojima (✉)
Daitobunka University, Tokyo, Japan

T. Shimaoka et al. (eds.), *Basic Studies in Environmental Knowledge, Technology, Evaluation, and Strategy*, DOI 10.1007/978-4-431-55819-4_3

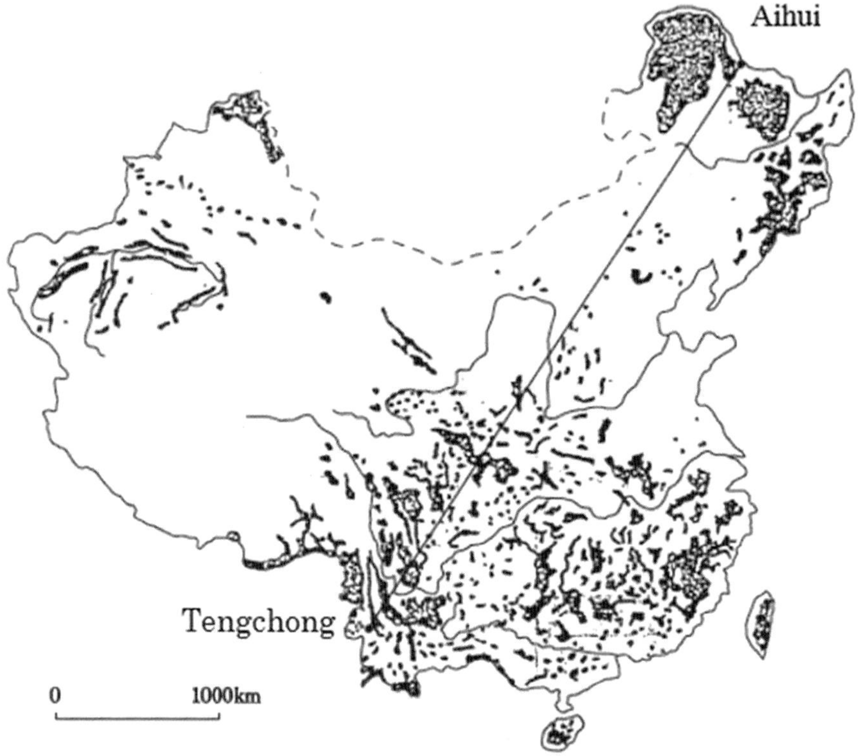

Fig. 3.1 Forest distribution map in the early 1950s (*Source*: Mu Zi 1956, p. 55)

Figure 3.1 shows a forest distribution map in the early days of nation building. Forests remained in Da Xing'an Ling and Xiao Xing'an Ling Mountains in the northeast, the upper basin of the Changjiang River, the hills in Fujian Province, and around the border of Yunnan Province in the south. Forests had disappeared in the agricultural area of the North China Plain, the Northeast China Plain, and the mid and lower basin of the Changjiang River. In particular, regarding this basin area, forests had disappeared to the extent of not being able to put colors on this map. Forest only accounted for approximately 8 % of the whole country area. This proportion is about the same standard as that found in desert countries. The northern tip of Aihui and Tengchong in the south is connected with a direct line. Ninety-four percent of the population lived on the eastern side of this line, while 6 % lived on the western side, mainly minorities.

Figure 3.2 shows the topsoil outflow distribution. The topography is roughly divided into three areas according to the height above sea level: plain (below 200 m ASL), hill (above that to 1500 m ASL), and plateau (1500–2000 m ASL). Topsoil outflow is the most serious on hills. It is estimated that there is an annual outflow of approximately 5 billion tons of topsoil. The soil ends up deposited on riverbeds which gradually become higher than the surrounding fields. This phenomenon is the

Fig. 3.2 Topsoil outflow (*Source*: China Environment Yearbook 1992 Edition, p. 20)

formation of a raised bed river. The water breaches the dike during periods of flooding, and it stays in the fields for a long time. This is called *laohai*, the greatest extent of damage caused by flooding. The mouth of the Yellow River protrudes into the Bohai Gulf, and this is a delta built up by soil deposits from the upper stream. Land is still being developed on the north side of Chongmingdao Island at the mouth of the Changjiang River.

Figure 3.3 shows salt damage distribution in a dry farming area where annual rainfall is 400–800 mm. Water evaporates from the topsoil after the fall harvest. Water in the soil is drawn up by capillary action, leaving basics on the surface. As a result, in January and February the farmland in the suburbs of Beijing and Tianjin becomes white as if it had been sprinkled with lime. After the fall harvest plowing has been done since ancient times in order to halt the capillary action. This is called fall plowing. The soil surface could be washed off if there were abundant water supply, but it cannot be done because of water shortage. This damage occurs on dry land and semidry land.

Figure 3.4 shows desertification in Inner Mongolia. No. 13 in the figure indicates ongoing desertification. Since the olden days, poor Han farmers would relocate to Inner Mongolia, Ningxia, and Qinghai. They cultivated the land and settled there while driving away the indigenous nomads. However, the number of such settlers

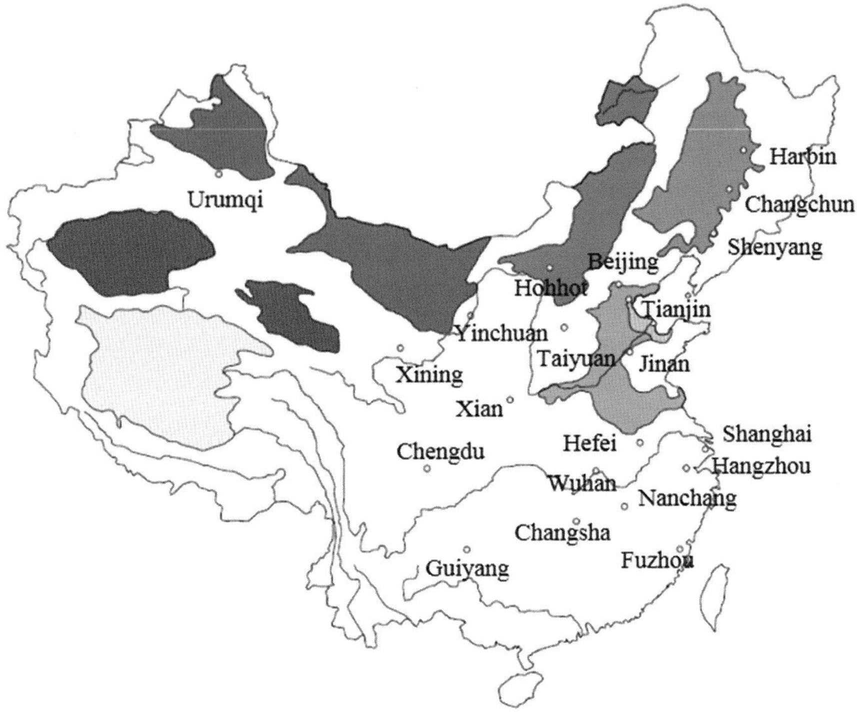

Fig. 3.3 Alkalization (*Source*: China Environment Yearbook 1992 Edition, p. 19)

was limited back then. In New China, the Communist Party systematically sent poor farmers and Liberation Army veterans there. An even worse thing happened in the Great Famine of 1960–1962 when a large number of Han farmers settled there and cultivated the land, harming the grassland. More and more *kohsa* yellow dust has been blown over to Japan since that time.

Those things are elements of the historical negative heritage of the ecosystem.

3.1.2 Nature of Industrialization Which Began Full Scale in 1953

The policy of industrialization has consistently emphasized heavy industry. In most developed countries, light industry preceded this. This system of development was changed by Stalin in the former Soviet Union who introduced a heavy industrialization policy, and China copied this policy. In reality, it was for developing war-related industries. To develop heavy industry, an incomparably larger amount of minerals as material and processing, consumed energy, in particular coal, and

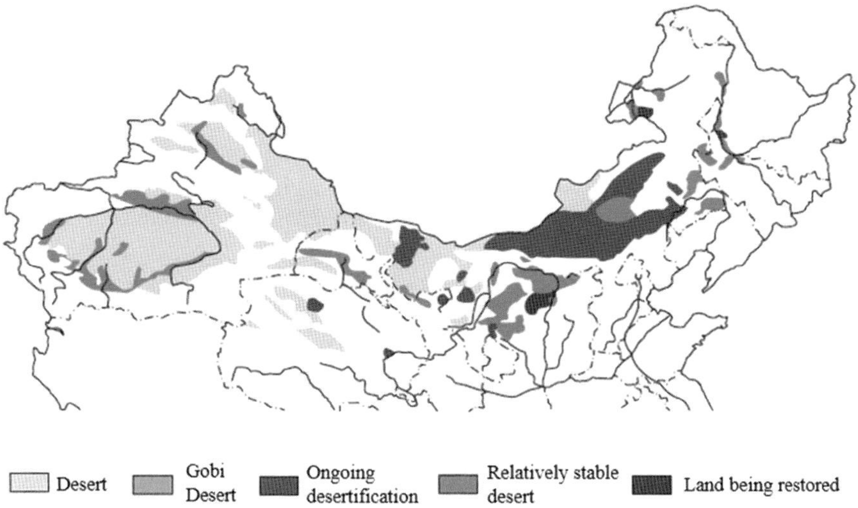

Fig. 3.4 Desertification (mid-1970s to mid-1980s) (*Source*: China Environment Yearbook 1992 Edition, p. 18)

waste are generated than they are in light industry. The environmental burden becomes quite considerable.

3.1.3 Occurrence of Compressed Environmental Destruction

Industrialization in developed countries took time, 100–200 years, while industrialization beginning later than the 1950s in developing countries took almost no time. In the case of developed countries, environmental destruction due to industrialization expanded from point to line, in certain areas, over large areas as much as covering the whole country, and on to global pollution. Pollution in a certain area occurred after WWII. It expanded to air pollution in entire cities due to dissemination of automobile use, while pollution spread in several villages or over the whole county due to aerial application of agricultural chemicals. Global pollution became apparent after the 1980s. In developing countries, these four stages tend to become compressed and occur at the same time, which makes it more complicated to come up with measures to deal with the pollution. Policies and investment for environmental protection need to have higher quality than those applied in developed countries.

3.2 1973 to the Mid-1990s Pollution Investigation and Policy Formation

Let us call this period the first stage of environmental protection policies. Environmental problems have five aspects: awareness of environmental pollution, investigating the actual situation, policy planning, establishing laws and environmental standards, and implementing policies. It is especially important to deepen the awareness of the government leaders who plan policies. There will be a huge difference in the degree of policy implementation depending on the protests of victims, converting awareness of leaders domestically due to accusation of experts or cool-headed investigation of actual situation, or urge for awakening to the awareness by external information or pressure. In the cases in China, under the Communist command economy system where priority was given to the government economic development policy, the press have been manipulated or suppressed by the authorities. Therefore, internal accusation has been limited. It was not until external information or a certain kind of pressure was applied when the leaders finally became aware of the situation.

3.2.1 Environmental Awareness

The United Nations Conference on the Human Environment was held in Stockholm in 1972. It was the first time that Chinese delegates had attended such a kind of international conference. From Japan, the patients and supporters of Minamata disease unofficially participated with a movie made by Mr. Noriaki Tsuchimoto showing the horrible condition of Minamata. Zhou Enlai, Premier of China at that time, immediately assembled National Environmental Convention upon reading the report of the delegates. This was the start of environmental administration, as related by Qu Geping who later became a central figure in environmental administration.

So was there no protest by the general public or accusation by experts? We still do not know about the former today. The latter did exist. In the "China Environment Yearbook" 1994 Edition when the ban on environmental reporting had been lifted, the editor published a paper titled an "Opinion concerning destruction of natural resources, reasonable use and protection in the future" which was submitted to the National Agricultural Engineering Scientific Conference in the spring of 1963 with a comment "An outstanding important report of 30 years ago by scientists." It had been written by Zhu Jifan and two others, signed by 25 experts. The paper pointed out the danger of environmentally destructive development and harm imposed on the land, biological creatures, marine resources, and mineral resources. Another piece was published in the *Guangming Daily* in 1965, an article concerning the problem of secondary alkalization. This newspaper is not a party organ of the Communist Party of China but one for intellectuals in the China Democratic

League. Both were written based on the Great Leap Forward of 1958–1960 during which hydraulic structure construction and primitive iron making (backyard steel furnaces) were widely developed in villages throughout the nation to an almost exclusive degree. The resulting agricultural neglect brought about a major crop failure, leaving the worst-ever case in history of 20–40 million people dead from starvation. This could eventually have led to criticism of Mao Zedong who was ultimately responsible for the disastrous policy, but the writer restrained himself to careful and cool-headed analysis. Such commentary could not become forceful enough to move the top-ranking government officials, for party considerations.

In addition, national forest inspection was conducted for 4 years from 1973 to 1976, which revealed that the amount of wood being felled was more than the growing amount of wood, that is, it was in deficit. This result reversed the forestry policy of the Forest Agency from felling trees to planting trees. However, forestry policy is almost powerless in the face of the whole nation's economic policies.

When it became possible to openly talk about environmental problems, He Naiwei published a paper "Danger of Changjiang River becoming second Yellow River (Huang He)" in 1978. In the following year in 1979, Cong Zimin disclosed a paper "Landfill on the lake shore requires immediate halting." He claimed that there were less than 500 lakes and ponds in the middle basin of the Changjiang River in Hubei Province, down more than half from the 1065 there had been before the nation building started. This was criticism toward the policy applied to reinforce food production. Furthermore, Wang Maoxiu published a paper "We must not become criminals in history" in 1980, an accusation of nature destruction by felling trees in the border area of southern Yunnan Province which used to be richly forested.

The Second National Environmental Convention was held in January 1984. There, the resolution "environmental problem is the fundamental national policy" was proclaimed. A little before that, the population problem had been determined to be the fundamental national policy, so environmental problem was the second fundamental national policy. It took 10 years from the first convention to unify the environmental awareness among the leaders. The several previously mentioned "accusing" papers seem to have contributed in their own way. Please note that all of those papers dealt with ecosystem destruction. It can be interpreted as a partial revelation that a historical negative heritage is no longer capable of tolerating industrialization.

3.2.2 Progress of the Inspections

Various inspections were encouraged in the following 10 years. Before those wide-ranging inspections, national research institutions had been carrying out local and partial inspections since around 1973. Inspections started on the mouth of the Yellow River, the mouth of the Pearl River, and the coastal area of Shandong Province. This was probably because of the heavy pollution comprised of visible materials such as floating waste and oil in the coastal area. They did not do research into the water quality at that time, but construction of observation stations began along the coast. Similarly, hydrological stations began to be reinforced along the main streams of major rivers. Some hydrological stations had existed since the period of national government, but they were basically inspecting the level of water in order to predict flooding. It was not until the 1990s when the stations were constructed along the branches.

A new field of inspection was in manufacturing, the key source of pollutant emission. During the 3 years from 1985 to 1987, 168,000 companies (36.4 %) among 460,000 state-run companies existing back then were inspected. The results were published in 1990. The key finding revealed that 65 % of industrial waste came from 3000 companies, 75 % came from 6000 companies, and 85 % came from 9000 companies. However, this inspection was limited to facilities in the manufacturing sector and covered only a little over one third. Moreover, they did not inspect other industries or aspects of domestic life.

Over 20 million township and village enterprises existed in the vast agricultural villages, where only experimental inspection was conducted. Fuel inspection was rather more important in agricultural villages. The agricultural population were prohibited from moving out of their villages right until the beginning of 1980 when the people's commune system collapsed. The increased population had no choice but to continue living in the villages where they were born. On the other hand, food was under a mandatory delivery system, and farmers could not retain enough rice for personal use. Fuel was only collected from the surrounding hills. The control of the people's commune began to become lax around 1976 or 1977. In the past, the whole community or village would cut down planted trees along the farm roads, sand preventive and windbreak trees, and trees along railroads, under the noses of government officials who turned a blind eye. Frequent orders to halt the tree cutting were being issued from the central government by the middle 1980s. The Ministry of Agriculture encouraged policies of small hydraulic power plants, small coal mines, and methane gas generating facilities. However, promoting small coal mine development later became a huge problem due to the resultant residue and combustion ashes.

Due to the progress of these inspections, the 1990 edition of the "China Environment Yearbook" was the first one to be officially published. Pollutant statistics began to be published at the end of the yearbook. The yearbook has been published annually with an increasing number of items in the statistic figures. There have been numerous books published in Japan on the environmental problems of China. Many depend on this attached statistics, and they are often not aware of the range of inspection in statistics for analysis. Analysis without constantly being aware of the range of inspection will invariably end up with misleading conclusions. The conclusion may be drawn that the amount of emission in narrow definition is decreasing.

Regulating laws and environmental standards will be omitted due to the limitation of pages.

3.2.3 Feature of Policies

Main actual policies up to the mid-1990s are listed here:

1. "Three-simultaneous system" policy
2. Environmental assessment system for newly developing project
3. "Anti-Pollution Levy" system
4. Responsibility system for making target of environmental protection by the companies
5. Responsibility system for making integrated indices of environmental protection by the mayor
6. Pollutant emission permit system
7. Pollutant centralized processing system
8. Complete processing system within the designated period at polluting factories
9. Closing order to factories

(1) and (2) are mandated upon new projects and new factory construction. (1) is that the environmental-related facilities should be incorporated at the same time of design, construction, and the beginning of operation of the main facility. (8) is applied for existing factories. (4) and (5) are systems imposed on the head of companies as well as mayors. GDP annual real growth rate has absolute importance among achievement assessment indices for mayors. For environmental field, multiple indices are indicated in score points, and the total points are used to assess achievement of the head of companies and mayors. Among the indices for which mayors are responsible, the GDP growth rate is vitally important. Mayors put all their efforts into increasing the growth rate; therefore, the environmental index is not very effective. (3) is a kind of tax where a certain ratio of tax is imposed on above standard concentrations when pollutants which are directly concerned with life are emitted, such as heavy metals. However, fieldwork revealed that some factory managers claimed they felt they were entitled to emit as much as they wanted, because they were paying the Anti-Pollution Levy, and it was more cost effective than building an attached processing facility. As for (7), it makes the situation more complicated to concentrate on one processing plant because Chinese companies would put branch factories with different waste matters in a plant area, and also there are a lot of housing developments for employees. (9) is a very drastic policy of imposing a closing order onto factories. However, this cannot be applied to large and mid-size factories because they are too big to fail. This policy is applied mainly to small factories, such as township and village enterprises in agricultural communities. However, when a closing order was placed on a factory, a new factory would be built on the other side of a mountain in the following month, and the operation would restart. This could happen any number of times, and the problem remained unsolved.

After all, these nine policies except (3) are intended to reduce emission by administrative order. Looking back on the experience in Japan after the 1960s, the effectiveness of any policy is lessened unless it is an economic policy where a building-attached environmental protection facility is eventually increasing company profit. There must be various ways to do it, for example, if the total amount of an environmental protection facility investment is included in a cost account or the depreciation period will be shortened if the total amount is not included in the cost. Furthermore, part of the expense is covered by a government grant. In Japan, even compost for domestic households is subject to a grant. In China, domestic waste is increasing every year, but no separation is done for collection.

What is more important than this economic policy is the fact that those who measure the emission and those who expose violation are both officials of the same Ministry of Environmental Protection. This means they are easily open to bribery. It is more or less customary to welcome government inspectors with the red-carpet treatment when they arrive and furnish them with gifts when they leave. In Japan, inspectors for government grant projects who are sent from the Board of Audit reject even the offer of a free lunch. The policies of certain administrative methods have a high potential to become a fertile breeding ground for corruption.

The press can be a watch dog. However, they are under the rule of the Communist Party; therefore, environmental destruction will imply not be reported unless the situation has become so bad that people have died. When the SARS outbreak occurred in 2003, right after the press conference by the Ministry of Health of State Council, an old doctor at a PLA General Hospital in Beijing leaked information to foreign press agencies that there had been more casualties from his hospital than reported in the public announcement by the Ministry of Health. He was prepared to be arrested. This was wired to foreign news agencies. After that, the world was astounded at the number of SARS casualties that were revealed. In rural areas, a very grave pollutant had made a great many people sick, but local party officials would not admit reporters. Foreign journalists were blocked from entry even more vigorously. Sometimes local government will hire thugs to deter reporters. Such stories are too numerous to mention, but in the end, antipollution public opinion is discouraged. Such lack of social devices significantly lowers the penetration of practical environmental policies even though the government is now making a big issue out of environmental protection.

3.3 Mid-1990s to Today: Environmental Burden Additionally Increased by Hyper-rapid Growth

Rapid growth does not quite express the economic development from the latter half of the 1990s to 2011. We need to add the prefix "hyper." Some critics even call this period the "Golden Age." There are two major factors: one is a major increase in exports bringing a vast inflow of foreign currency and the other is an extraordinary rise in investment for infrastructure and real estate development. These created a new gigantic burden on the environment. The situation is shown in Fig. 3.5.

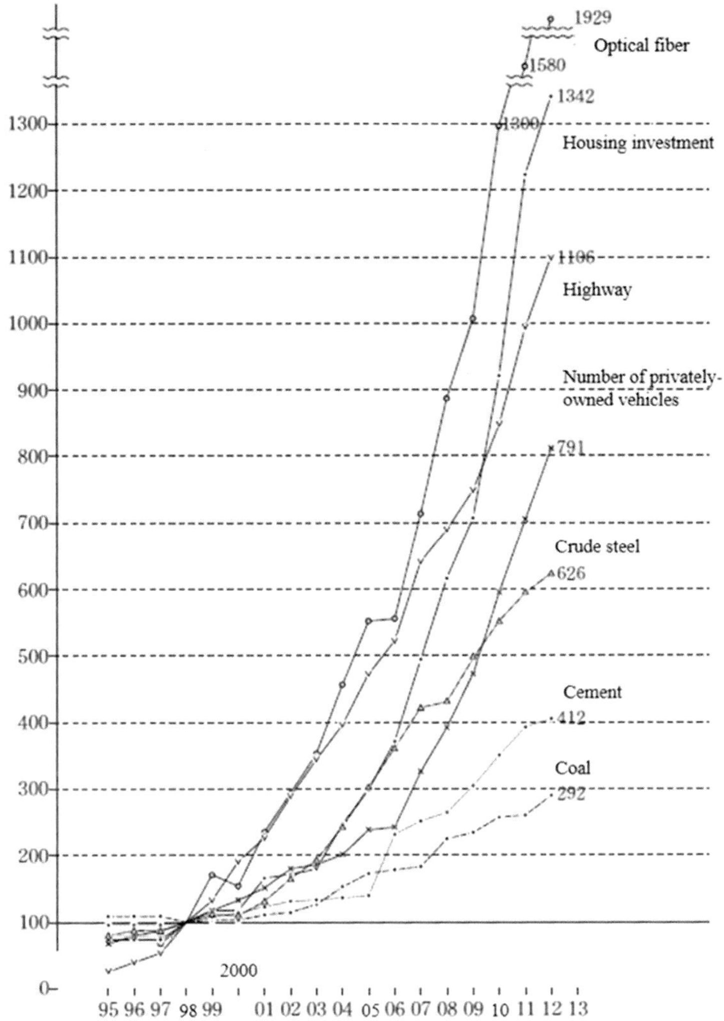

Fig. 3.5 Big jump in the number of related industries emitting waste on a large scale (1998 = 100)

3.3.1 *New Environmental Burden Brought About by Qualitative Change in the Economy*

The following qualitative changes were brought about by the hyper-rapid growth during this period: the first is the plunge into an era of mass production, mass consumption, and mass disposal, the second is the huge development of material industries and related mining, and the third is the significant expansion of coal development due to a rapid increase in energy consumption.

As can be seen from the figure, showing indices concerning infrastructure and real estate development, the total distance covered by paved highway increased by 8 times, the investment for housing increased by 13.4 times, and the amount of optical fiber grew by 19 times, all in 14 years. In material industries, cement production grew by four times and crude steel grew by over six times. In 2012, China produced nearly 60 % of the cement in the whole world and 48 % of the crude steel. The growth of coal in this country was 2.8 times more. In actual amount, production was 3.5 billion tons. Supplemented by 1.8 billion tons of imported coal, it is almost 3.7 billion tons. How much sulfur and incinerated ash would be generated when such a vast amount of coal is burned? It is too scary just to calculate it. The influence of investment in infrastructure and real estate development on related industries is particularly large. Material industry induces a large amount of material production, encouraging mine development and transportation in order to transport minerals. When a mine is developed, a large area of land surface is stripped, and a large amount of slag is deposited. Similarly, a material processing department discharges a large amount of slag. The number of privately owned vehicles reached 150 million. If military vehicles are included, the country has twice as many cars as Japan. In Japan, more than one third of these are so-called light vehicles, a phenomenon not widespread in China. This may be because of the flashy consumer activity of larger and more expensive cars selling better than small ones, since for a long time vehicles have been a symbol of social status. Such vehicles emit more exhaust gas than necessary.

As for the penetration of mass production, mass consumption, and mass disposal in rural agricultural villages, probably we should grasp it in a wider view of a relative modernization wave in such places. In consumption, washing machines, refrigerators, and motorcycles are owned by over two thirds of households, and there are two mobile phones for every household. There are more than 1.2 times the number of TVs than there are households. We can say new lifestyle has been disseminated into almost two thirds of all households. From the viewpoint of environmental burden, this new lifestyle creates a throwaway culture. Added to the total urban population, we can consider 80 % of Chinese people are now enjoying such a lifestyle.

What is important in rural villages from the viewpoint of environmental problems is mechanization and house rebuilding. Due to mechanization in agriculture, unclaimed parts of harvest such as shells and husks, which had been recycled into the fields as compost for thousands of years, were now discarded and burnt. This began happening in the mid-1990s. Urban residents claim that this smoke is a source of pollution. The Ministry of Agriculture issues orders to recycle the plant waste into the farmland, but in reality things are happening in reverse. Recently remote-sensing photographs are being taken via man-made satellite for warning where burning is happening. This works toward weakening the soil fertility.

When migrant working becomes common and farmers save money, they always build a new house in their hometown. In many houses, the traditional bathroom is converted to new style, and human waste is no longer used in the agricultural fields. The elements of the Chinese character for "house" consist of a pig under a roof. A pigsty would traditionally be made in a part of a house where pigs were fed with food scraps and human waste. Swine manure would be recycled into farmland, creating an ideal recycling economy. In the southern coastal and delta areas, toilets were made above fish-breeding ponds where dropped human waste was consumed by fish, another recycling system. This system is disappearing as well. Only half of consumed human food is absorbed by the body. Human waste was "food with less nutrition." Recycling human waste had built an agricultural system with high productivity per unit of land since ancient times. In the last 10 years or so, the system of raising stock has changed drastically. Large-scale poultry farms, hog farms, and cattle farms are built in the suburbs of cities, which becomes the mainstream. Feces exceeding the capacity of treatment go into rivers. The total amount of feces would reach approximately 4 billion tons estimated from the population and number of farm animals. It is estimated that 70 % of this flows into rivers. Human waste in cities stopped being recycled into farmland in the early 1980s when the expansion of the urban area increased the distance for transporting waste. This phenomenon began spreading to rural villages at the end of 1990s. Feces are liquid and are heavy. You have to apply it only in the farmland around the house in order to do so looking at the harvest.

As a result, agriculture increasingly became dependent on chemical fertilizer and agricultural chemicals. Soil pollution occurred as a result. Furthermore, the demand for bricks has increased as construction material for new housing. Soil was removed from farmland to be used as brick materials. Banning orders for this have been issued numerous times.

Another factor contributing to pollution of rivers in agricultural villages is the development of agricultural processing companies using agricultural harvest as materials. Wastewater from breweries, farm animal products processing companies, leather industries, and paper mills is increasing significantly.

As is seen here, all this new development is creating new environmental burden.

3.3.2 Deepened Environmental Awareness Accompanying New Economic Change

I would like to explain two important opportunities for this. The first was the United Nations Conference on the Human Environment held in Rio de Janeiro in 1992. Premier Li Peng attended this conference as the representative. The "Framework Convention on Climate Change" and "Agenda 21" were adopted at this conference. Besides these, the global consensus on the management, conservation, and sustainable development of all types of forests (forest principles) was a nonlegally binding authoritative statement. I believe this was the first United Nations conference when the importance of forests was focused on for preventing global warming. It seems the Chinese leader acknowledged the huge importance of restoring greenery to deal with environmental destruction. The Chinese version of Agenda 21 was made and published in 1994, which emphasized the importance of ecosystem maintenance.

Another important factor is the result of large-scale research on the ecosystem in China which was published under the title of "The research of land resource throughput and population carrying quantity in China" in 1991. It is a thick book of 1551 pages. The expression "population carrying quantity" in the title means "capacity to feed the population." The research was conducted in cultivated areas, mountains and forest, and water resources. It was designed in 1986 when China was still afraid of population increase and was not confident regarding food self-sufficiency. What I was intrigued by was the viewpoint of how much population could be supported in a vessel containing forest and water. There is an interesting point in the result of research report. The cultivated area expanded by 37 % from what had been disclosed. The mandatory delivery system was installed in 1953, and the population began to increase substantially after 1964. Farmers could live only in people's communes. How could the increased population survive? It came to light that in fact there were concealed rice fields.

The United Nations Conference on the Human Environment where environmental problems are globally dealt with for the whole earth and the domestic Chinese thought of environmental problems, which should be thought of as being in the vessel of population feeding capacity of cultivated land, forest, and water, became quite close to each other in the late 1990s.

3.3.3 Groping for New Policy

Fieldwork on the environment is the most difficult of all types of research due to the wide coverage of range and diversity in the research items. Therefore, it is inevitable to focus on the key pollution area or the pollution which directly involves human lives such as heavy metal pollution. The five major areas are listed below:

1. Pollution: Three rivers and three lakes.

   Three rivers:

   Huai River – the source is in Shandong Province and Anhui Province. It flows through the plains in Jiangsu Province to the East China Sea.
   Liao River – the source is in Liaoning Province. It forms the Northeast China Plain in Liaoning Province.
   Hai River – the source is in Hebei Province. It flows through Beijing and Tianjin before entering the Bohai Sea.

   Three lakes:

   Lake Tai – a large lake neighboring Jiangsu Province and Zhejiang Province. It also serves as a water source for Shanghai city.

Lake Tian – a lake located to the south of Kunming, the capital of Yunnan Province.

Lake Chao – a lake located to the south of Hefei, the capital of Anhui Province.

The basin of the three rivers and three lakes covers 810,000 km^2 (an area 2.2 times as large as Japan), spreading over 14 provinces and direct-controlled municipalities with a population of 360 million.

2. Areas of serious air pollution: areas with serious SOx pollution. Pollution is very heavy in these areas consisting of 27 province-level administrative bodies and 175 cities. The total area is 1.09 million km^2 (three times as large as Japan).
3. Key environmental projects by the central government.

   Three Gorges Dam Project

   South-to-North Water Diversion Project – a waterway from the Han River, a large tributary of the Changjiang River, to Tianjin, covering a distance of approx. 1200 km
4. One city: Beijing city
5. One sea: Bohai Sea

These areas are designated as the most focused areas where control at the source and total volume control method are introduced. This method was applied at the Seto Inland Sea for water pollution control and the Keihin Industrial District for air pollution control in Japan. These focused areas in China cover a vast range. Beijing city alone is as large as the total area of Tokyo, Kanagawa, and Saitama prefectures, the western half of Chiba prefecture, and the eastern part of Yamanashi prefecture. There is an anecdote to give one an idea of the size of China. One of the delegates of the Li Hongzhang party representing Qing state visiting Shimonoseki at the end of the Sino-Japanese War commented that he had heard there were no rivers in Japan. However, he said that he had found one, which turned out to be the Seto Inland Sea. Chinese people live in a place where the sense of area is extraordinarily large, so the focused areas are large as well. What we should note is that these are the areas with the most serious pollution. Other areas are not accounted for. Research projects are conducted in these focused areas.

As for monitoring air pollution, PM2.5 and O_3 are added to the monitoring items in major cities in 2011. This was attributed to external pressure. The U.S. Embassy was monitoring air quality in their garden. The resultant data was sometimes revealed, and there was hearsay evidence that their data was worse than that from the monitoring data of Beijing city. It hurt the most vulnerable part of China.

In the second stage, monitoring inspection advanced steadily and it has become precise. That being said, we should realize that the actual situation of the agricultural village area remains untouched where the problems are the most serious. Wu Xiaoqing, the Vice Minister of the State Environmental Protection Administration (SEPA), made a remark at the greeting of the National Natural Ecosystem Protection Conference held in June 2006: “According to an *estimate*, (1) annual wastewater in agricultural villages in China amounts to over 8 billion tons, domestic waste is 120 million tons, and animal waste is 2.5 billion tons. There is still a problem with drinking water for 300 million people. 'Soiled water is overflowing in various places, waste is littered here and there, brushwood and grass is randomly left untouched, and animal feces are scattered everywhere' and this is the general situation in the villages today. (2) The situation of industrial waste from the cities dumped in agricultural villages is becoming worse. Agricultural villages in the suburb of cities have become disposal sites for urban domestic waste and industrial waste. The lost farmland amounts to 2 million mu (134,000 ha). A particularly pressing problem is waste from small and mid-sized companies in agricultural villages. The polluted area spreads far and wide, but it is difficult to monitor, manage and control them rigorously. It is even more difficult to remove this accumulation. Incidents among the people due to this situation are increasing every year. (3) Overall soil pollution is serious. The harm is on a massive scale, having a major impact on food safety and endangering human health. It does not only cause a direct economic loss but also endangers the safety of ecosystems in the country. (4) Ecosystem destruction in agricultural villages is even more serious. Large scale mining of a large amount of soil and rocks, collection of river sand, removal of soil from farmland to make bricks, land reclamation in lakes and ponds, and development in the hills all cause ecosystem destruction." ("China Environment Yearbook" 2008 Edition, p 75)

I highlighted the word *estimate* at the beginning of the paragraph. The figures in the statistics themselves are estimates. They are not the research results. Moreover, animal feces amount to 3 billion tons in the "China Environment Yearbook" 2013 Edition. There are no statistics for human waste. This citing is more enough to reveal the serious situation in agricultural villages.

3.3.4 Groping for a New Environmental Policy

As mentioned already, environmental protection policies which have been in place since the 1980s are basically in the administrative order style which has numerous loopholes. Economic policies to replace them are still at the stage of groping in the dark. Pan Yue, the Vice Minister of the State Environmental Protection Administration (SEPA), talked about economic method they were groping for at a conference held in September 2007 ("China Environment Yearbook" 2008 Edition, pp. 64–68): (1) Introduction of an environmental tax, in particular one levied on companies with high energy consumption, high resources consumption, and which generate high pollution. An internal list is already drawn up for the chemical industry. (2) Raising the existing "Anti-Pollution Levy." As was previously described, he points out that the current amount is too low, and it tends to encourage discharge into the environment more than its intended purpose. (3) Low-interest loans to environment-friendly companies. (4) Ecosystem compensation system. Compensation is required for those which caused damage to the ecosystem. (5) Opening a market for "Anti-Pollution Levy right." This is similar to a market for emission rights of carbon dioxide. (6) Revoking tax refund for exporting products whose energy consumption/resources consumption is high in trade policy. (7) Creating an environmental insurance system.

Among these seven items, items two and three will be in the experimental stage within 1 year, and the experiment will be complete within 2 years. Full-scale operation is targeted within 4 years. Meanwhile, he also says "Policy making is very slow and is not progressing well. The biggest reason is adjustment of authority and interest among various ministries in the State Council, various industries and local governments. When conflict arises among the three parties, the State Environmental Protection Administration (SEPA) does not have the authority. It is 'easier said than done'." The administration holds authority below ministry level. Five years later, in the "China Environment Yearbook," the 2013 Edition includes no article which mentions introducing even one of the above seven items. The highest leaders deliver impressive speeches at international conferences, but we can see they are cold toward agencies which are in charge of actual and important fields. The cult of self-righteousness pervading the ministries still holds sway.

3.3.5 Investment for Environmental Protection

From an economic viewpoint, control of environmental protection depends on investment for protection. So, investment for environmental protection statistics is examined. The summary is shown in Table 3.1.

There is no clear definition of investment for environmental protection. As for China, p. 441 of the "China Environmental Statistics Reference Materials 1981–1990" lists, "the total sum of expenditure of each item in pollution measures which were budgeted and executed in the fiscal year by companies or non-profit organizations." There is no telling if expenses for restoring the ecosystem are included or not. Since it was the time when only the environmental problems in a narrow sense were the object for environmental problems, it is highly probable they were not included. Since the investment for this part is negligible even today, it will not cause many problems for international comparison.

Table 3.1 Change of investment for the environment

China		Comparable country		
Year	Ratio to nominal GDP (%)	Country	Year	Ratio to GDP (%)
1981–1985	0.56	Netherlands	1991	2.43
1986–1990	0.69	Japan	1990	2.27
1991–1995	0.73	Switzerland	1993	2.18
1996–2000	1.06	Germany	1991	2.13
2001–2005	1.2	Sweden	1991	1.57
2006–2010	1.4	Finland	1994	1.36
2011–2015 (estimated)	1.2–1.4	France	1994	1.23
		U.K.	1994	1.23

Source: China: 1981–2000: Kojima (2000 p. 55)

The figure for 2006–2010 is 2.16 trillion Chinese yuan, figures of 5 years according to "China Environment Yearbook" 2013 Edition, p. 45, divided by the sum of nominal GDP in the same 5 years. The figure for 2011–2015 is the estimate investment amount of 3.4 trillion Chinese yuan in p. 19 of the said Yearbook divided by the estimated GDP growth rate of 12 %. Comparable countries: "Study on Establishing Satellite System for Integrated Environmental and Economic Accounting" The Japan Research Institute, Limited, 1999. GDP in this book is not limited to nominal or real. Usually nominal GDP is used for denominator to calculate this tax for precision.

Looking at this Table, we can see that the ratio is rising every year, yet it is still very low compared with the figure for western countries and Japan. The Western countries and Japan have been making investments at or above this standard for the past 30–40 years. As a result, the current environmental standard has been achieved. China has finally become closer to low standard countries since about 2006. As was already described in 1 about the peculiarity of China, there are existing situations where more environmental destruction will happen than had happened in developed countries. When this is taken into consideration, China needs to raise the ratio to around 3 % of nominal GDP and to maintain the standard for 30–40 years. Otherwise the deterioration will not be halted. When the restoration of the ecosystem, which has a negative historical heritage, is taken into consideration, I feel around 5 % must be invested every year for 30–40 years. The mindset of the leaders, to augment military expenditure alone, is a thoroughly misguided notion.

3.4 Assessment of the Effect of a 40-Year Period of Environmental Protection Activity

Kuznets once oriented his hypothesis on economic development and restoration of environmental destruction from the viewpoint of economics. The ecosystem was outside his viewpoint. The rate of increase of pollutant waste is higher than the development rate of the economy for some time after economic takeoff begins. Then the reverse situation becomes the norm at a certain point. This is attributed to the increase in investment for environmental protection, development of pollution reduction technology, and awakening of people's awareness concerning environmental destruction. Some refer to this as the Kuznets Turning Point.

Whether the 40 years of effort since 1973 to prevent environmental destruction in China is about to reach this Turning Point or not, the author did hold the view, until several years ago, that China was reaching this Turning Point, at least in some items. For example, when the emission statistics of each pollutant attached with the "China Environment Yearbook" are minutely examined, we find that the maximum of emission was in the 1980s or 1990s and emission was decreasing in the twenty-first century, in some items. However, recently I have come to think that these statistics are correct but wrong. It is because this method of estimating the un-researched part from the part subjected to research and the results create more disparity as the coverage widens. Moreover, we should always interpret publicly disclosed statistics with a degree of reserve. That is why the research scope has often been referred to in this paper. Emission sources are unlimitedly vast in the agricultural villages, in industries other than manufacturing, and various types of domestic waste, all significantly lowering the credibility of statistics. When modernization comes, traditionally useful things become waste. Human waste is a typical example. It is even more difficult to grasp the actual situation of ecosystem destruction from statistical figures.

It is difficult to apply Kuznets hypothesis for judgment as a whole in consideration for this point. Therefore, publicly disclosed figures of pollution, destruction, recycling, etc., are collected and indicated with the three symbols of ↗, →, and ↘ for improved, no change, and worsened (Table 3.2). The basis of publicly disclosed figures has been omitted due to limitation of pages.

Table 3.2 Assessment for each item of environmental problems

		Items of environmental destruction	↗ Improvement → No change ↘ Worsened
Environmental waste in narrow sense	Air	Suspended particulate matter	↗
		PM2.5, O_3	↘
		SOx	→
		CO_2	↘
		NOx	↘
	Water	Heavy metals	↗
		COD	↘
		Domestic wastewater	↘
		Others	↘
	Solid waste	Discharged amount	↘
		Total usage amount	↗
		Domestic waste	↘
Ecosystem		Water shortage	↘
		Land subsidence	↘
		Underground water level	↘
		Soil pollution	↘
		Loss or destruction of cultivated land	↘
		Planted and natural forest area	↗
		Grassland	↘
		Topsoil outflow	↘
		Alkalization	→
		Desertification	↘
		Biodiversity	↗

The State Planning Commission at the time of 1976 issued instructions to make long-term plans for environmental protection. The principle was "suppressing pollution in 5 years and resolving it in 10 years." Twenty years later, the environmental plan in the Ninth Five-Year Plan aimed at "stopping worsening of environmental pollution and ecosystem by 2000 and converting the ecosystem from worsening to restoration by 2010." The results as of 2013 are shown in the Table. The understanding of the complication of environmental problems and the difficulty in gaining solutions remain still overly optimistic and full of wishful thinking. They are basically just mouthing beautiful words with little or no substance to them.

3.5 Conclusion

First: The leaders in China lacked awareness of environmental destruction accompanying economic development till 1972. The awareness stemmed from international pressure to change it, not protest movement of victims or analysis by experts in China.

Second: The awareness deepened due to a serious shortage of farmland, water, and forest, as well as international environmental awareness. It was the awareness that people can survive only in the vessel of the whole resources of the nation.

Third: Policies against environmental destruction are mainly administrative orders and not based on economic methodology. Administrative orders are simply not very effective.

Fourth: Policy penetration is weakened by the dictatorship system suppressing the will of the people.

Fifth: Investment for environmental protection is increasing, but it is still not enough considering the size of environmental pollution and ecosystem destruction caused by too-rapid economic development.

Sixth: The environmental protection activity has gone on more than 40 years and is substantial. However, little effect has yet been seen due to the scale of the pollution problem. The coming 20–30 years will be the most crucial period. Probably, China has finally reached the stage of entering into suppressing deterioration and improving the situation.

Those are the key points of this paper.

One addition is China's comments on global environmental problems and attitude toward international pollution originating in China and affecting surrounding countries. China has claimed at international conferences that the main responsibility for global environmental pollution should be placed on developed countries and such countries should provide environmental technology free of charge. Surely the claim of "free" was removed in the twenty-first century, but it still maintains the attitude of responsibility on the developed country theory, even though 20 % of the world's population is emitting 25 % of the CO_2. Moreover, solid waste and wastewater from Mainland China are polluting the nearby ocean. The country is a major source of dirty air and is polluting the air of neighboring countries. Not only that, it is reducing the amount of water flowing into South Asian and Southeast Asian countries downstream by building numerous dams on the rivers in the continent. The Tibet highland is an important water resource to supplement the water shortage of China. It maintains the arrogant stance of keeping its mouth shut regarding this situation.

References

China Environment Yearbook (1992) China environment yearbook 1992 edition

Kojima R (2000) Structural change in modern China – will environment restrict growth? University of Tokyo Press, Tokyo

Mu Zi (eds) (1956) Knowledge of forest. Center for Chinese Research Materials, Association of Research Libraries

Chapter 4
Environmental Issues in Southeast Asia

Michikazu Kojima

Abstract This region has continued to experience economic growth since the 1980s, driven by the expansion of direct investment from other countries. However, this growth has also brought with it a host of environmental issues, including deforestation, air pollution, and water pollution. This chapter provides an overview of the economic status, environmental regulations, and other aspects of Southeast Asia, which require a coordinated response amid economic integration, including those that encompass several countries, such as cross-border air pollution. It discusses the initiatives that have been implemented so far and examines the remaining challenges of dealing with regional-level environmental issues.

Keywords Southeast Asia • Haze • Biodiversity • Mekong River • Trade and the environment

4.1 Introduction

Southeast Asia comprises the 11 countries of Brunei, Cambodia, East Timor, Indonesia, Laos, Malaysia, Myanmar, the Philippines, Singapore, Thailand, and Vietnam. The per capita GDP (purchasing power parity) among these countries varies considerably, from USD 70,000 in Singapore, a very high level, to USD 1,200 (or lower) in East Timor.

Despite these economic disparities, the Association of Southeast Asian Nations (ASEAN) was established in 1967 and has been gradually moving toward economic integration. Integration is also proceeding in social aspects, including politics and the environment. Currently, East Timor is the only Southeast Asian nation excluded from ASEAN.

This region has continued to experience economic growth since the 1980s, driven by the expansion of direct investment from other countries. However, this growth has also brought with it a host of environmental issues, including deforestation, air pollution, and water pollution. In its report summarizing economic development in ASEAN countries through to 2030, and related issues, the Asian

M. Kojima (✉)
Institute of Developing Economies, Japan External Trade Organization, Tokyo, Japan
e-mail: michikazu_kojima@ide.go.jp

T. Shimaoka et al. (eds.), *Basic Studies in Environmental Knowledge, Technology, Evaluation, and Strategy*, DOI 10.1007/978-4-431-55819-4_4

Development Bank Institute highlighted environmental and resource issues (Asian Development Bank Institute 2014). Some of the challenges highlighted for sustainable development include the prevention of air pollution and water pollution, energy issues, the sustainable use of renewable resources such as forests, and the protection of biodiversity.

In this book, Chaps. 1, 2, and 3, as well as 5, discuss the individual environmental issues and environmental policies in the Asian countries. This chapter provides an overview of the economic status, environmental regulations, and other aspects of Southeast Asia, in the next section. I then discuss the environmental problems that require a coordinated response amid economic integration, including those that encompass several countries, such as cross-border air pollution. Here, I introduce the initiatives that have been implemented so far and examine the remaining challenges of dealing with regional-level environmental issues.

4.2 Diverse Southeast Asian Countries and Their Environmental Issues

The Southeast Asian countries differ significantly from one another in terms of income levels and industrial structure. Table 4.1 shows the population, per capita GDP, manufacturing sector share of GDP, and governance indices of each country. Indonesia accounts for 40 % of the population of Southeast Asia, with 240 million people. The Philippines and Vietnam each have a population of approximately 90 million, and Thailand approximately 70 million. Then, Singapore has by far the highest per capita GDP, with a level of purchasing power parity higher than that of Japan. In addition, Malaysia's per capita GDP is approximately USD 16,000, a level that admits membership to the OECD, the so-called club of advanced nations. The country with the highest manufacturing sector share of GDP is Thailand (36 %), while Malaysia and the Philippines stand at 24 %.

Generally, as a country's manufacturing sector develops and per capita income increases, environmental issues, such as water and air pollution, emerge, prompting the implementation of pollution regulations. Table 4.2 summarizes the legal system relating to the environment in each country. Singapore, Malaysia, Thailand, Indonesia, and Vietnam appear to have basic legal systems in place, but there are no adequate systems in the other countries.

Moreover, even if regulations are in place, in situations in which there is insufficient monitoring or rampant corruption, pollution regulations are not properly enforced. The governance indices highlight whether the government is effective and corruption is under control. In this case, Singapore has a high general governance capacity. In countries with low per capita income levels, corruption is generally evaluated as not being under control. Therefore, there is a high possibility that, even if regulations exist, they will not be functioning effectively. Furthermore, another condition for advancing pollution countermeasures is that victims of

Table 4.1 Southeast Asian countries macro indices

	Population (2012, millions)	GDP (2012, million US dollars)	Per capita GDP (2012, US dollars, PPP)	Government effectiveness (2012)	Regulatory quality (2012)	Corruption control governance index (2012)
Singapore	5.3	286,908	75,913	2.15	1.96	2.15
Brunei	0.4	16,953	72,917	0.83	1.16	0.64
Malaysia	29.2	305,032	22,280	1.01	0.55	0.30
Thailand	66.7	365,965	13,976	0.21	0.23	–0.34
Indonesia	246.8	876,719	9,009	–0.29	–0.28	–0.66
Philippines	96.7	250,182	6,109	0.08	–0.06	–0.58
Vietnam	88.7	155,820	4,998	−1.53	−0.68	−0.56
Laos	6.6	9,386	4,464	−0.88	−0.84	−1.04
Cambodia	14.8	14,054	2,839	−0.83	−0.35	−1.04
Myanmar	52.7	n.a.	n.a.	−0.29	−1.87	−1.12
East Timor	1.1	1,355	1,179	−1.19	−1.02	−0.98

Source: World Bank *World Development Indicators* website and *World Governance Index* website
Note 1: *PPP* purchase power parity. Using exchange rates based on price level

Table 4.2 Environmental laws in Southeast Asian countries

	Basic law	Air pollution control	Water pollution control	Waste management	Environmental impact assessment
Singapore	○	○	○	○	
Brunei					
Malaysia	○	○	○	○	○
Thailand	○	○	○	○	○
Indonesia	○	○	○	○	○
Philippines	○	○	○	○	○
Vietnam	○	○	○	○	○
Laos	○		○		
Cambodia		○	○	○	
Myanmar	○				
East Timor	○				

pollution speak out and that their voices are recognized as important political issues. The economic and social conditions relating to the enforcement of pollution regulations are further aspects in which diversity is evident among the Southeast Asian nations.

Table 4.3 is an environmental performance index collated by Yale University. Twenty indices are used, covering the nine fields of health impacts, air quality, water sanitation, water resources, agriculture, forest, fisheries, biodiversity and habitat, and climate and energy. By assigning the best performing countries on

Table 4.3 Environmental performance index (2014)

	Overall ranking	Overall	Air quality	Water and sanitation	Water resource	Forests	Biodiversity and habitat
Singapore	4	81.78	98.33	100	99.65		46.33
Brunei	37	66.49	94.62	73.93	37.84	36.46	100
Malaysia	51	59.31	90.54	77.21	8.64	1.68	93.37
Thailand	78	52.83	67.67	57.62	16.00	25.34	70.19
Indonesia	112	44.36	75.31	24.29	0.02	7.75	78.08
Philippines	114	44.02	81.53	37.35	0.53	31.35	64.67
Vietnam	136	38.17	51.32	43.15	0.14	17.25	43.39
Laos	127	40.37	29.23	17.5	0	13.28	93.85
Cambodia	145	35.44	64.8	10.52	0	0	78.93
Myanmar	164	27.44	47.68	30.69	0	24.47	28.62
East Timor	132	39.41	69.33	12.22	0.3	31.35	60.43

Source: Yale University *Environmental Performance Index* http://epi.yale.edu/

the relative index a score of 100, and the worst performing countries a score of 0, the different indices are integrated into a single index. Table 4.3 adds the five indices of air quality, water sanitation, water resources, forest, and biodiversity and habitat to the integrated index and gives a ranking score among 178 countries. The highest scoring Southeast Asian country on the integrated index is Singapore, which ranks fourth. The lowest scoring country is Myanmar, which ranks 164th. Environmental issues and countermeasures in Southeast Asia are often mentioned in the context of a single region, but in fact, they vary considerably between the countries.

4.3 Environmental and Resource Issues and International Cooperation

In Southeast Asia, there is a need for capacity development to implement environmental countermeasures for each country. At the same time, many problems require the relevant countries to cooperate, including cross-border environmental pollution, international river water use, and the maintenance of ecosystems. Other areas will require systems to be aligned as economic integration proceeds, such as chemical substance regulations, energy-saving standards for products, and recycling systems. This section gives an overview of these issues and introduces the kinds of initiatives that have been promoted in Southeast Asia.

4.3.1 Mekong River

The Mekong River is an international river originating in China, and the river basin spans Myanmar, Laos, Thailand, Cambodia, and Vietnam. The development of shipping and hydropower using the Mekong River was first proposed just after the Second World War, but various international conflicts and other issues meant there was little real progress until the mid-1990s. In 1991, the civil war in Cambodia ended, and the focus returned gradually to the development of the Mekong basin. However, concerns have arisen over the impact of dams and other developments on ecosystems and human livelihood, water shortages in downstream regions, and so forth.

In 1995, Thailand, Laos, Cambodia, and Vietnam concluded a four-nation agreement, the Agreement on the Cooperation for the Sustainable Development of the Mekong River Basin (Mekong Agreement), establishing the Mekong River Commission. The Mekong Agreement define the notification and consent scheme, depending on the main river, tributaries, and rainy and dry seasons, with respect to use within the basin and division of water outside the basin. Three cases are indicated: cases requiring the commission be notified, cases in which prior consultation must be held to form a consensus in the commission, and cases in which the agreement of the commission must be obtained (see Table 4.4). Water usage that requires notification, prior consultation, and agreement includes irrigation, hydropower, navigation, flood control, fisheries, timber floating, recreation, and tourism, thereby covering a wide range of river usage. The item garnering the strongest interest in connection with environmental issues is dam development. According to the provisions of the Mekong Agreement, dams built on the main river require international consultation prior to construction, while for dams built on tributaries, notification alone is sufficient.

Table 4.4 Stipulations on notification and agreement on Mekong River usage in basin and water division outside of basin

	Mekong River tributaries including the Tonlé Sap		Mekong River main river	
	Use in the river basin	Water division outside the basin	Use in the river basin	Water division outside the basin
Rainy season	Notification to the commission		Notification to the commission	Prior discussion by the commission aimed at forming a consensus
Dry season			Prior discussion by the commission aimed at forming a consensus	Consensus in the commission

Source: Created based on Article 5 of the *Agreement on the Cooperation for the Sustainable Development of the Mekong River Basin*

There are concerns over the impacts of dam construction, such as the destruction of ecosystems and the losses of people whose livelihoods depend on ecosystems. Ziv et al. (2012) analyzed the impact of existing and planned dams on water resources in the lower Mekong basin (the so-called Golden Triangle, bordering Thailand, Laos, and Myanmar), including 11 dams on the main river and 78 dams on tributaries. They conducted a scenario analysis, starting with a baseline in 2015, that included 41 tributary dams and 0 main river dams. In their analysis, they compare a scenario in which all 78 dams planned on tributaries are built by 2030 and a scenario in which no tributary dams are built, but 6 dams are built on the main river. As a result, they estimate that water resources would be reduced by 19.1 % in the former case and 6.9 % in the latter case. On the other hand, they also estimate that the amount of electricity supplied would increase by 24 TWh/year in the former case and 35.8 TWh/year in the latter.

However, a study focusing on dam construction in the lower Mekong basin alone does not entail a comprehensive study of the impacts on the Mekong River and its ecosystems. The reason is that dam construction is also occurring upstream, in China. Matsumoto (2005) and International Rivers (2013) pointed out that dam construction in China's Yunnan Province may have potentially negative impacts on the Mekong Delta. First, the amount of earth and sand flowing to the lower Mekong basin may decrease. Second, the water level during the dry season may increase, preventing the riverbank cultivation practiced during the dry season. Third, there are potential issues surrounding a sudden rise or fall in water levels.

One of the problems with the Mekong River Commission is that the upstream countries, China and Myanmar, do not participate, although dialog between these two countries and the Commission does occur. In 1996, there were talks between the Mekong River Commission and Myanmar and China. In 2002, an agreement was reached with China in which the commission would receive flow data and water level data, collected at two locations in China. The background to this development was a flood of the Mekong River in 2000 (MRC website: http://www.mrcmekong.org). Nevertheless, there is no framework for evaluating the environmental impact of dams constructed in China and Myanmar or discussing these with countries in the lower basin. Thus, the agreement could not function sufficiently from the perspective of comprehensive management of the Mekong River basin.

4.3.2 Forest Fires and Smoke Pollution (Haze)

Major forest fires in Indonesia have a history stretching back to 1982–1983 (Gillert 1998). In that forest fire, 3.2 million hectares of forest burned in eastern Kalimantan. Thereafter, forest fires occurred in 1989, 1991, 1994–1995, and 1997–1998. Forest fires release massive quantities of air pollutants into the atmosphere, causing haze. In recent years, forest fires have been occurring every year, and the resulting haze is becoming an issue. The impacts of haze are wide ranging, including respiratory and eye diseases associated with atmospheric pollution, problems in transport systems such as delays in airline schedules and flight cancelations, a decline in the number of tourists, and so on. The resulting damage from these fires affects the country where the forest fire occurred, as well as neighboring countries such as Singapore and Malaysia. Glover and Jessup (1999) estimated the damage from forest fires and haze in 1997 to have reached USD 4.5 billion. Of this, damage caused directly by the forest fire (e.g., damage to agriculture and forestry) amounted to USD 3.1 billion, and damage to short-term health and to the tourism sector amounted to USD 1.4 billion. Damage inside Indonesia was estimated to be USD 3.8 billion, while the combined damage to Singapore and Malaysia was estimated to be USD 0.7 billion.

Forest fires have also been indicated as a significant contributor to global warming. Silvius et al. (2006) estimated that CO_2 emissions in 1997, 1998, and 2002, when major forest fires occurred, ranged from 3.0 billion to 9.4 billion tons. Here, 9.4 billion tons corresponds to 40 % of the global emissions of CO_2.

Large-scale forest fires are attributed primarily to natural factors and human. Natural factors include a decline in rainfall due to weather fluctuations such as El Niño or the existence of subterranean peat layers. Human factors include slash-and-burn cultivation, dehydration of forests due to forest thinning, and burn-off associated with converting forest into agricultural land for the development of palm plantations or rice paddies. In recent years, the burn-off for palm plantation development has been attracting particular criticism.

Initiatives at ASEAN entered full swing in 1997. The ASEAN Ministerial Meeting on the Environment formulated the Regional Haze Action Plan. Then, in 1999, the Regional Haze Action Plan Coordination and Support Unit responsible for haze was established within the ASEAN Secretariat. In 2002, the ASEAN Agreement on Transboundary Haze Pollution was finalized and signed. By 2003, the agreement had been ratified by six countries, Malaysia, Singapore, Brunei, Vietnam, Myanmar, and Thailand, and was later ratified by Laos, Cambodia, and the Philippines. In 2014, Indonesia finally ratified the agreement. Currently all of ASEAN countries have ratified the agreement.

The agreement defines haze as "smoke resulting from land and/or forest fire which causes deleterious effects of such a nature as to endanger human health, harm living resources and ecosystems and material property and impair or interfere with amenities and other legitimate uses of the environment." Moreover, "land and/or forest fires" are defined as "fires such as coal seam fires, peat fires, and plantation fires."

Signatory countries are required to mandate the prevention of land and/or forest fires in law. For example, Section 9a requires signatory countries to promote a zero-burning policy.

> *Developing and implementing legislative and other regulatory measures, as well as programmes and strategies to promote zero burning policy to deal with land and/or forest fires resulting in transboundary haze pollution.*

In 2003, the guideline for the Zero Burning Policy was formulated (ASEAN Secretariat 2003). Tacconi et al. (2008) point out that it is not necessary to stop all burn-off, as stipulated in the ASEAN Agreement on Transboundary Haze Pollution.

Instead, it would be better to focus on preventing burn-off of peat layers. It seems there is a need to perform a verification analysis to determine which activities need to be stopped and to then implement relevant countermeasures. Furthermore, Wetland International (2006) note out that it is important to restore wetlands at low cost.

Various initiatives are proceeding in parallel with the haze agreement. The ASEAN Peatland Management Initiative has been operational since 2003 and focuses on peatland by raising awareness and promoting sustainable peatland management. The ASEAN Peatland Management Strategy 2006–2020 was created in 2005 and updated in 2013 (ASEAN 2013).

In May 2012, the ASEAN-Wide Fire Danger Rating System was established. Indices such as humidity and ease of fire spread were created and published on the ASEAN Haze Action Online website (http://haze.asean.org/). The ASEAN Specialized Meteorological Center was also established, with its website publishing satellite image information, such as areas where a forest fire is occurring and/or haze is occurring, as well as the wind direction in the respective areas.

Forest fires in Indonesia have been studied and monitored for some time, but in 2011, the Sub-Regional Ministerial Steering Committee on Transboundary Haze Pollution in the Mekong Sub-Region was established and held its first meeting. This was more in response to reports of haze occurring in regions bordering the three countries of Thailand, Laos, and Myanmar, such as northern Thailand and northeastern Myanmar.

Despite the various initiatives that have been implemented, the issue of forest fires and haze has failed to reach a resolution after nearly 20 years. The key to forest fire countermeasures going forward will be moving beyond establishing international frameworks, including companies developing palm plantations in approaches for sustainable development. Furthermore, there is a need to look beyond central governments to find ways to involve local governments and communities in creating an effective framework.

4.3.3 Initiatives Related to Biodiversity Protection

Countries in Southeast Asia participate in international frameworks that protect biodiversity, such as ratifying the Convention on International Trade in Endangered Species of Wild Fauna and Flora (CITES) and other biodiversity protection agreements. At the same time, they are also developing a framework for regional cooperation aimed at protecting biodiversity in Southeast Asian countries.

The Coral Triangle Initiative on Coral Reefs, Fisheries, and Food Security program was established in 2007 to respond to the crisis concerning coral reefs and marine resources.[1] The participants include four countries from Southeast

[1] http://www.coraltriangleinitiative.org/

Asia—Indonesia, Malaysia, the Philippines, and East Timor—as well as Papua New Guinea and the Solomon Islands. The initiative has its secretariat in Jakarta. The Coral Triangle region accounts for only 1.6 % of the planet's ocean area, but includes some 600 types of coral, corresponding to 76 % of known varieties, and 2,500 species of coral-dwelling fish or 37 % of known varieties. However, excessive harvesting of marine products and pollution from land are believed to be putting the ecosystem at risk.[2] In response, training manuals have been created, and capacity-building programs are being implemented that target regional governments.

The Heart of Borneo Initiative is a three-nation initiative by Indonesia, Malaysia, and Brunei to protect biodiversity on the island of Borneo.[3] In 2007, the countries designated an area called the Heart of Borneo (HoB) as a biodiversity preservation initiative. In 2009, the HoB Strategic Action Plan was formulated. The action plan calls for a variety of capacity-building activities related to biodiversity, collaboration on research into the sustainable development of the HoB by forming links with research institutions, raising awareness to prevent further loss of forest biodiversity, and other such programs. The total area of Borneo is approximately 725,000 km^2, of which 200,000 km^2 has been designated as the HoB.

The ASEAN Center for Biodiversity was formally established in 2005 as a continuation of the ASEAN Regional Center for Biodiversity Conservation project, which had run from 1999 with support from the European Commission. The center is headquartered in the Philippines and holds workshops on topics such as biodiversity in urban areas, climate change, and biodiversity. It also performs GAP analyses related to marine reserves.

4.3.4 Cross-Border Products and Environmental Regulations on Products

In Southeast Asian countries, various environmental regulations now apply to products when shipping them to markets. This is an attempt to control global warming and chemical pollution and is based on regulations in advanced countries, such as those in Europe.

Environmental product regulations related to energy saving include Minimum Energy Performance Standard (MEPS) as minimum performance standards for each product and energy-saving labels to show a product's energy-saving performance in a multistep fashion. The MEPS has spread gradually from the Philippines in 1993 to Indonesia in 2004. Energy-saving labels are an ongoing initiative, with

[2] Refer to the Coral Triangle Initiative website (http://www.coraltriangleinitiative.org/).

[3] Refer to the Heart of Borneo Initiative website (http://www.hobgreeneconomy.org/en/about/).

Thailand being the first to achieve the No. 5 energy-saving label.[4] The initiative started in 1995. Then, from around 2005, other Southeast Asian countries strengthened their energy-saving label initiatives, with energy-saving labels being introduced in Malaysia (2005), Singapore (2008), and Vietnam (2013). However, the initiative has many discrepancies, including different methods of measuring energy-saving performance and different reference values used for labels.

Regulations on lead, cadmium, and other chemical substances included in electrical products were introduced in Thailand in 2009 and in Vietnam in 2012. Thailand uses a similar approach to Japan in that they require labels showing the degree to which chemical substances are included in a product, as compared to a designated standard. Vietnam uses a system similar to the European RoHS system, requiring designated substances to be below a certain standard. However, when different regulations have to be applied to the same products, then the products must comply with the respective markets, increasing the burden on manufacturers (Michida 2014).

Southeast Asian countries are grappling with issues related to urban waste, such as insufficient collection services, pollution from landfill processing sites, and difficulties in situating landfill processing sites. In response, there are moves to introduce extended producer responsibility, obliging manufacturers and importers to collect and recycle used products in the same way as in advanced countries. In Vietnam, a prime ministerial decision, announced in August 2013, will see extended producer responsibility applied from 2015 to 2018 on used products such as batteries, electrical products, lubricant oil, tires, motorcycles, and cars. New product categories will be added in stages. In Malaysia, a regulation is being prepared to extend producer responsibility of discarded electronics and electrical equipment. Indonesia is planning to promote autonomous business initiatives while phasing in extended producer responsibility over a 10-year period. As the cross-border movement of people and products increases while, at the same time, there are different systems in place for applying extended producer responsibility, situations in which the responsibility for used products is not clear will become increasingly likely.

In applying the abovementioned energy-saving standards, chemical substance regulations, and extended producer responsibility, there has, thus far, been no systematic coordination between Southeast Asian countries. Energy-saving standards and chemical substance regulations can place unnecessary economic burdens on producers, for example, in product design and calculations. In addition, extended producer responsibility certainly has the potential to create systemic distortions when used products move across borders and producers must fulfill their responsibilities. Therefore, there is a need for better coordination among ASEAN countries, including adopting unified standards.

[4] The Philippines was the first country in Southeast Asia to establish an energy-saving law. The initiative began in the early 1980s, but with the collapse of the Marcos government (1986), the initiative ceased to operate.

4.4 Southeast Asian Economic Integration and Environmental Issues

There are plans to liberalize trade among ASEAN countries toward the end of 2015. Various measures are planned to promote economic integration, including opening the markets of the service industry sector by allowing people to move more freely and so forth.

As this kind of economic integration proceeds, there is a possibility that factories will relocate to areas where environmental regulations are weaker, and investments that cause environmental problems may increase. The problems of haze and the management of the Mekong River basin, introduced in Sect. 4.2, have both been affected by economic integration in Southeast Asia. With regard to the haze issue, there is significant investment by Malaysia and Singapore in palm plantation development in Indonesia. Water transport on the Mekong River supports part of the trade between countries in the Mekong River basin. Moreover, electricity generated by hydropower stations in the Mekong River basin, particularly in Laos, is sold to other Southeast Asian countries. Increased economic integration carries the potential to make these problems even more serious.

The EU and NAFTA have adopted countermeasures to ensure that environmental problems do not grow more serious as economic integration progresses. The United States, Canada, and Mexico concluded the North American Agreement on Environmental Cooperation at the same time as they concluded the NAFTA. To ensure that environmental issues do not grow worse with the liberalization of trade, a commission was established comprising the ministers for the environment of the three countries (Commission for Environmental Cooperation). Moreover, the secretariat for the commission prepares public announcements of environmental information and environmental terminology in English, French, and Spanish. The EU has established an organization called the Implementation and Enforcement of Environmental Law. Initiatives are now underway to maintain a unified standard of enforcement of environmental regulations by undertaking capacity building for countries that subsequently joined the EU and by conducting joint investigations into the status of cross-border movement of waste products (Kojima and Michida 2011).

In ASEAN, official and unofficial meetings of the environmental ministers and high-level summit meetings have been held every year, such as the first ASEAN Ministerial Meeting on the Environment in 1981. The Blueprint adopted in 2007 calls for the creation of joint bodies, not only for economic issues but also for political, health and safety, and social and cultural issues. Environmental issues are positioned within the ASEAN Social-Cultural Community (ASCC). The Blueprint presents the following 11 priority items in a section on Ensuring Environmental Sustainability:

1. *Addressing global environmental issues*
2. *Managing and preventing transboundary environmental pollution (transboundary haze pollution and transboundary movement of hazardous wastes)*
3. *Promoting sustainable development through environmental education and public participation*

4. *Promoting environmentally sound technology*
5. *Promoting quality living standards in ASEAN cities/urban areas*
6. *Harmonizing environmental policies and databases*
7. *Promoting the sustainable use of coastal and marine environment*
8. *Promoting sustainable management of natural resources and biodiversity*
9. *Promoting the sustainability of freshwater resources*
10. *Responding to climate change and addressing its impacts*
11. *Promoting sustainable forest management*

The progress on the Blueprint from 2009 to 2012 was summarized in a midterm review (ASEAN Secretariat 2014), which detailed initiatives in each of the above areas. However, there was no detailed discussion of areas such as those given in Sect. 4.3 Moreover, the report states that the various effects of the initiatives take time to become apparent and that there was no clear trend to indicate whether the environment had improved between 2009 and 2012.

Of the themes highlighted in Sect. 4.3, product environmental regulations and recycling systems are not included in the aforementioned 11 priority items. However, in the high-level summit on the environment for ASEAN held in January 2013, the notion of sustainable production and consumption was gaining importance, there was a proposal to expand for "environmentally sound technology" (ASEAN Secretariat 2014) to include "Sustainable Production and Consumption." Therefore, it seems likely that this will become an ongoing ASEAN initiative.

Ideally, when various environmental countermeasures are promoted, each country should bear a fair portion of the cost and receive a benefit. However, as noted in Sect. 4.1, the different countries in Southeast Asia have very different levels of income, economic scale, environmental laws, and so forth. There is a need for each country to bear costs in line with its capability to do so and to provide appropriate support to those countries with weaker environmental management capabilities. In addition, donors from outside the region also play an important role.

Furthermore, based on the agreement of each country, there is a need to increase the capacity of government and research institutions with respect to gathering environment-related statistical data and the status of policies. Much of the research cited here has been carried out by researchers outside of the region. In particular, with cross-border environmental issues with a certain level of apparent consensus in scientific knowledge, it is difficult to get all the countries to cooperate on promoting countermeasures. There is a need to foster exchanges between the researchers of Southeast Asian countries, to have them share scientific knowledge, and to promote discussion among scientists and policymakers from each country.

4.5 Conclusion

Economic growth and the advance of economic integration in Southeast Asian countries are beginning to cause cross-border and environmental and resource issues. Moreover, the introduction of local environmental product regulations by each country could place an overly heavy burden on companies supplying products within the ASEAN region as the economies continue to integrate.

On the other hand, with regard to haze, Mekong River management, and biodiversity protection, agreements have been formed and frameworks established for environmental protection and resource management. However, not all relevant countries are participants in these frameworks, including countries from outside the region. For this reason, activities based on international frameworks are limited to those such as raising awareness and sharing information.

To address these issues, the EU and NAFTA initiatives described in Sect. 4.4 offer a useful model. ASEAN countries need to implement more comprehensive initiatives and to cooperate in dealing with environmental and resource issues in the region.

The basis of comprehensive initiatives is the accumulation of scientific knowledge. It is important for countries in the region to promote efforts to grasp the actual status of environmental and resource issues and to study and share countermeasure technologies.

References

ASEAN Secretariat (2003) Guidelines for the implementation of the ASEAN Policy on zero burning. ASEAN Secretariat, Jakarta

ASEAN Secretariat (2013) Strategies and action plan for sustainable management of peatlands in ASEAN Member States 2006–2020. ASEAN Secretariat, Jakarta

ASEAN Secretariat (2014) Mid-term review of the ASEAN socio-cultural community blueprint (2009–2015). ASEAN Secretariat, Jakarta

Asian Development Bank Institute (2014) ASEAN 2030: toward a borderless economic community. Asian Development Bank Institute, Tokyo

Gellert PK (1998) Brief history of Indonesia's forest fire crisis. Indonesia no. 65, pp 63–85

Glover D, Timothy J (1999) Indonesia's fires and haze: the cost of catastrophe. Institute of Southeast Asian Studies and International Development Research Center, Singapore

International Rivers (2013) Lanchang river dams: threatening the flow fo the lower Mekong. http://www.internationalrivers.org/files/attached-files/ir_lacang_dams_2013_5.pdf

Kojima M, Etsuyo M (2011) Trade and environment. In: Masahisa F, Ikuo K, Satoru K (eds) The economics of East Asian Integration: a comprehensive introduction to regional issues. Edward Elgar, Cheltenham, pp 479–500

Matsumoto S (2005) The Mekong region: incorporating the views of regional civil society. In: Japan Environmental Council (ed) The state of the environment in Asia. Springer, Tokyo, pp 131–148

Michida E (2014) Policy impact of product-related environmental regulations in Asia. IDE discussion paper no. 451

Silvius M, Kaat A, van de Bund H, Hooijer A (2006) Peatland degradation fuels climate change. Wetlands International, Wageningen

Tacconi L, Jotzo F, Quentin Grafton R (2008) Local causes, regional co-operation and global financing for environmental problems: the case of Southeast Asian Haze pollution. Int Environ Agreements Law Econ 8(1):1–16

Ziv G, Baran E, Nam S, Rodriguez-Iturbe I, Levin SA (2012) Trading-off fish biodiversity, food security, and hydropower in the Mekong River Basin. Proc Natl Acad Sci U S A 109 (15):5609–5614

Chapter 5
Environmental Policies in East Asia: Origins, Development, and Future

Akihisa Mori

Abstract Environmental policies vary, depending on how a country frames and defines environmental challenges. Underlying causes of the environmental challenges can be classified as market failure, undefined ownership, government and/or institution failure, and globalization and their combination. The government of each country has seen the underlying causes as they wanted and chose policy instruments based upon their recognition. Coupled with the differences in economic development, pressures to the environmental challenges, and their management capacity, this has brought about difference in the choice of policy instruments, enforcement, effectiveness, and distributional impacts.

For an environmental policy to be more effective, it is indispensable for the government to frame environmental challenges and to address their underlying causes properly. Then all the government ministries and the political leaders should share the proper framing and definition so that they will take the environment into account in their sectoral policies, in other words, implement preventive measures and convince people and firm to integrate the environment into their activities.

Keywords Framing • Underlying causes • Economic growth • East Asia

5.1 Framing Environmental Challenges and Environmental Policies

Environmental policies are put in place by governments to address environmental deterioration that is officially recognized as being a problem. Therefore, policies differ, depending on how the government frames and defines environmental deterioration.

The government does not always define the problem that enables it to properly address the underlying causes. Sometimes the problem is defined to make the results appealing to the citizens or to expand the interest of the political leaders and their supporters.

A. Mori (✉)
Graduate School of Global Environmental Studies, Kyoto University, Kyoto, Japan
e-mail: mori.akihisa.2a@kyoto-u.ac.jp

T. Shimaoka et al. (eds.), *Basic Studies in Environmental Knowledge, Technology, Evaluation, and Strategy*, DOI 10.1007/978-4-431-55819-4_5

In case deforestation is viewed as excessive logging without consideration for the recovery of the forest, logging ban or allowing it within an ecological limit becomes appropriate policy response. However, such a policy may pose a political risk if the logging company exerts political influence. The political party can avoid such a risk if the government defines the underlying cause as shifting cultivation (slash-and-burn farming) by farmers living in the forest: as long as farmers who are operating shifting cultivation have feeble political and economic influence and thus have difficulty in delivering their voices, the government regards banning shifting cultivation and providing an alternative livelihood (which is unfavorable for farmers) as the "appropriate" policy measure. The weaker the farmers are, the more likely the government defines the underlying cause like this.

On the other hand, in case the government defines the underlying cause as a lack of government capacity, it becomes convincing for the forest department to implement measures that expand their interests, say, to increase their staff and budget.

Referring to the abovementioned relationship between framing and "appropriate" solution, this chapter takes a stock of how East Asian countries have defined underlying causes of environmental challenges and implemented environmental policies. Then we prospect development of their environmental policies.

5.2 Framing Environmental Problems in Neoclassical Economics

Neoclassical economics has framed environmental problems as market failure due to environmental externality or lack of market. People tend to undervalue the environment, thus put inappropriate price in the market. This leads to excess consumption of environmental goods and services, forcing society to bear the cost of environmental deterioration or bring about resource depletion.

Unclear definition of property right provides additional incentive for the excess consumption. So long as the ownership of environmental resources is properly defined, the owner is expected to engage in sustainable use in order to maximize their value and to dissuade illegal development/use of environmental resources.

In reality, however, the government does not define legally binding property right to all the environmental resources. It does not admit legal registration and define ownership of some lands, especially ones located in frontier areas where the central government is hard to control, while land is one of the most basic components of human economic activity. In Japan, Toyotomi Hideyoshi, the governor in the late 1500s, made a comprehensive survey to define land ownership and productivity as a way to identify the taxpayer and amount of tax levied (Taiko kenchi). The survey was limited to paddy rice field and did not cover mountains and forests. In the seventeenth century, the successor defines most of the mountains yielding abundant timber as the ownership of the Tokugawa shogunate. However, communities nearby were allowed to manage the areas as common-pool resources and to harvest non-timber forest products such as firewood, charcoal, or bamboo shoots to supplement their income. It was not until the Civil Code was established in 1891 during the Meiji era that the representative of the community became the owner of the land on the consensus that the land was commonly owned (owned by the community). While the radical land reform after the World War II kept the land ownership in the forest areas intact, inheritance tax charged to the registered representative de facto accelerated the shift from communal to private ownership. Coupled with significant decline in earning from forest and increasing income opportunity at urban and industrial areas, heavy inheritance tax has prompted children of the owners to give up land ownership and immigrate to urban areas. Insufficient inspection and updating of local government have increased the unclearly defined land ownership in those areas.

Furthermore, there remains uncertainty in predicting environmental impacts that current economic activity may cause. Humans do not completely know the natural laws governing the ecological system. This disables them to have a future market to make an inter-temporal trade of environmental resources. Cost of environmental resource depletion and deteriorating environment in the future are not fully reflected in the price in the current market.

Globalization can expand the abovementioned environmental externality. Freer international trade and foreign direct investments may not only expand resource depletion and industrial development that will cause environmental pollution but also increase consumption that brings about greater waste emission. In the long term, a country may specialize into industries with high resource and/or environmental intensity in order to make the best use of comparative advantage in foreign trade. This will simply aggravate resource depletion and environmental pollution in resource-rich and capital-rich economies.

When we frame environmental challenges like this, the solution will be to legally define property right to environmental resources. This will constitute the basis for appropriate valuation of the environment. The market will bring about an efficient solution if the environmental value is appropriately priced as a tax or in the emissions trading.

On the other hand, environmental problems can be framed as policy and institutional failure. A government implements policies for economic development in order to attain multiple purposes of poverty alleviation, enhancing legitimacy of the ruling party, and increase in revenue for their discretion. It also implements policies to protect and expand the vested interest of their political supporters such as the military, political elites, industry, and massive population. While it implements policies and initiates development projects that benign for society and environment, reducing regional and/or urban-rural income gap pushes the ones that may cause negative environmental impacts. In addition it often disregards voices of the affected people. The government even takes over land entitlement that local communities have traditionally and collectively used to give concession on logging and/or reforestation for private companies and to protect watershed. It sometimes does this without agreement of the communities, excusing that such entitlement has not been legally defined.

An environmental challenge can be also recognized as the "implementation deficit"; that is, even if the government department in charge of environment implements more stringent environmental laws and regulation, other ministries and local governments do not enforce them strictly and companies do not comply with them. This holds especially true in cases of weak penalty for violation, of difficulty in proving cause and effect, or of high transaction costs incurred to reach an agreement.

These explanations imply that environmental problems will not be solved by government initiative if the main underlying cause is the policy or institutional failure. Even if international society convinces the government to implement environmental policies that are comparable to them, there is no guarantee that the government is willing to and is capable of implementing them strictly. It is worth scrutinizing political economy of environmental policies, especially domestic factors that work to develop them.

5.3 Economic and Institutional Cause of Environmental Problems in East Asia

East Asian countries have suffered from a variety of environmental challenges, which can be framed as the one of market failure, unclearly defined ownership, government and/or institutional failure, and globalization.

First of all, they have taken development strategy and promoted compressed industrialization that enabled them to rapidly catch up with the industrialization process that developed countries had gone through over a century. They have placed higher priority on economic growth than environmental conservation. They have not developed environmental policies and institutions in commensurate with the industrialization. They were not willing to take policy instruments that shift the environmental costs to polluters that caused environmental pollution and destruction at the earlier stage of economic development as they are the main drivers of industrialization. That is to say, the government left the market failure untouched. This holds true not only in capitalist countries but also in socialist countries where the government adjusted production and consumption and thus was expected to control environmental pollution under a planned economy. In reality, socialist countries did not operate a planned economy in such a way that could control pollution. For example, China dispersed industrial plants to a variety of region, not for the sake of minimizing pollution but for fear of possible future invasion of the United States. However, this location dispersion disabled China to take advantage of economies of scale, lowering resource efficiency, and worsening environmental pollution.

Allocation of property right was also an underlying cause of environmental degradation. Land is defined as state ownership not only in socialist countries such as China and Vietnam but also in capitalist countries such as the Philippines and Thailand. Only land use title is allowed to private companies and residents. The government provided logging concession to private companies as a means of acquiring foreign exchange, disregarding traditional entitlement of the communities. However, private companies often conducted illegal logging beyond the boundaries as the government had limited capacity to monitor their activities. They overexploited the environmental resources while did not fulfill their duty of reforestation. Farmers also encroached the degraded forest for agricultural extension. Such activities caused frequent flooding and increased tensions with downstream farmers and urban dwellers. This came to the forest policy change from production to conservation. Nonetheless, this policy change has shown mixed results in terms of conservation. In the Philippines, on the one hand, the government allowed communities to grow up economic trees under the reforestation program, reconciling community livelihood with forest conservation. The Chinese government launched a sloping land conservation program in 1998 that forced farmers in the government-designated reforestation areas to plant ecological as well as economic trees, in exchange of 5–7 years of compensation for their income loss. In Thailand, on the other hand, the government has promoted industrial plantation of nonnative but fast-grown and economical tree species such as *Eucalyptus* and *Acacia* in the name of reforestation program, and it raised a concern that this had negative impacts on the ecological system.

In order to tame the frustration of poor people, the government has supplied utility services at a subsidized price. It has not frequently adjusted the price in accordance with inflation to stabilize consumer price. This resulted in excess consumption of environmental resources at the expense of government budget. Such government price subsidization is greater in resource-producing countries. In Indonesia, for example, the share of fuel subsidy in government expenditure raises up to 18.8 % in 2005 and it rose as high as 22.6 % in 2008, increasing budget deficit. The Indonesian government decided to cut the subsidy in 2005 and 2008. It suspended subsidy cut in 2012 in the face of citizens' protests, but it was implemented in 2013 when it was predicted that the United States would abolish its purchase program of Indonesian national bonds. However, it turns out to be a temporally decline in demand as rising incomes increase demand for cars and accordingly transport fuel. China, by contrast, turned from a net energy exporter to an importer in the late 1990s and gradually cut fuel subsidy and liberalized coal market to allow price increase in the late 2000s.

Impact of globalization was greater among developing countries that attracted massive foreign direct investment for export-oriented growth while did not develop capacity for the environment. In Thailand, capital-intensive petrochemical industries became densely located, causing serious environmental pollution in eastern seaboard of Map Ta Phut. Globally increasing demand, coupled with local governments' provision of concession without strict enforcement, has drawn massive investment on palm oil plantation, accelerating deforestation and causing haze in ASEAN region.

5.4 Experience of Environmental Policies in East Asia

At the early stage of economic development, governments in many East Asian countries regarded that victims should accept environmental damages for the sake of nation or society in general (Mori 2012a). They unheeded complaints and petitions that victims made against industrial pollution and development projects. They tried to tame the victims by paying small amount of compensation or financial assistance without investigating the cause and effect or altering the production technologies or designs of development projects. These measures did not stop environmental deterioration and made the damage worse. Victims stepped up protest against polluters and developers, resorting to force such as blocking the transport of goods to industrial plants or the project sites. In countries that established a democratic institution such as Japan, Korea, Taiwan, and Thailand, victims organized themselves to build and develop a large-scale environmental movement. They brought environmental cases to the court and supported candidates who were proclaiming environmental improvement to be elected as governors, mayors, and parliament to place pressures for policy change.

It was not until these radical conflicts, lawsuits, and elections forced companies and governments to pay compensation and to suspend plants operation or construction that the government recognized industrial pollution as being a bottleneck on economic development. Then they enacted environmental laws and regulations and organized environmental administrative and fiscal mechanisms at central and local governments. They also prescribed public hearings and participation in environmental impact assessment laws and constitutions, attempting to make environmental impact assessment a practical means of consensus building among stakeholders.

In Japan, the government defined prevention of human health damages as environmental policy goal after several years of serious health damages from industrial pollution that was spread over Japan. Amid the enthusiasm that supported priority in environmental protection over economic growth, the government came to see stringent industrial pollution policies as indispensable means for sustainable economic growth and proclaimed the policy goal that reduced environmental pollution to the level of not causing health damage. The government initiated epidemiological surveys to set environmental standard below which health damage is not caused. It also accelerated research and development of pollution abatement technologies that ensured to achieve the environmental standards, demonstrating to urge companies to invest on them.

These countries did not frame industrial pollution as market failure. The government initiated to develop environmental infrastructure including sewage systems and waste treatment/disposal facilities to accept both industrial and domestic wastewater/waste to clean up the dirty environment within a short period. It went further to provide subsidized loans with companies that invested on pollution abatement technologies even if these policy measures were against the polluter pays principle.

Some governments even regard pollution haven and/or resource depletion in neighboring low-income countries as a "solution," instead of tackling domestic environmental challenges seriously. The Thai government has faced fierce protests against new coal-fired power plants since Mae Moh lignite-fired power plant discharged massive sulfur dioxides that led to hospital admission of more than a thousand nearby residents. After this incident, the 1992 Environment Act and the 1997 Constitution clearly stipulated environmental impact assessment (EIA) prior to the government approval of development projects. Nonetheless, the government approved the Hua Hin coal-fired power plant project prior to EIA. This made people lose credibility to the government. The government did not implement industrial pollution prevention policies that were stringent enough to eliminate serious health damage at the Map Ta Phut Industrial Estate in spite of almost 20 years of frequent and fierce protests. With the support of environmental victims in other region, the local residents filed a lawsuit in the administrative court at last. In 2009, the Supreme Administrative Court ruled that 65 out of 76 new industrial plant projects in the Map Ta Phut had to be suspended for the reasons of serious health damage among local residents and flaws in the procedure concerning negligent legislation stipulated in the Article 67, Section 2 of the 2007 Constitution. While 14 projects got approval by March 2010 and 25 projects could go ahead, this court ruling made it difficult to construct petrochemical industrial complex. Nonetheless, worsening political turmoil and social division after the 2006 coup disabled the Thai government to implement more stringent environmental regulations or to make more strict enforcement (Mori et al. 2010). This has prompted the Thai government to resort to financial assistance for, or participating in, hydroelectric power generation projects in Laos and Myanmar as well as natural gas development and industrial estate development projects in Myanmar as a way of mitigating bottleneck on economic growth (Mori 2012b).

China is also recognizing environmental pollution as a bottleneck on economic growth. It is going through serious environmental pollution, causing health and ecological damages and land exploitation around the nation, raising a number of protests against factories, development projects, and waste incineration plants. The Chinese government frames the underlying causes as obsolete state enterprises and local protectionism; that is, the local government does not strictly enforce environmental laws and regulations on local companies to protect them and thus local government vested interests (Kitagawa 2011). Based on this framing, it implemented stringent environmental regulations and increased environmental investment, coupled with the transition to the market economy to make large and key state enterprises survive while shut down low competitive ones. This framing, however, has led the government to see vehicle as the underlying cause of PM 2.5 instead of industrial plants and quality of transport fuel, allowing the China National Petroleum Corporation, one of the major state enterprises in China, to resist to the government requirements on upgrading investments. This has made it difficult to enhance effectiveness of countermeasures against air pollution.

In order to overcome the local protectionism, the State Environmental Protection Administration (SEPA) initiated shutdown polluting plants in the designated heavy pollution areas such as the Huai River basin, allowed watchdog functions to citizens and media, and disclosed environmental rating of the firms as a way of complementing government-limited environmental capacity, but to the extent they did not cause fierce opposition movement against the government. It has also implemented the total emission control on sulfur dioxide and COD in the environmental five-year plan since 1995 and allocated a reduction target to each local government to comply with them. To ensure their compliance, it implemented a "one-vote rejection system" that includes achievement of energy saving and/or total pollutant emission targets as one of the key performance evaluation criteria of local government leaders. Finally, it has provided supporting measures, financial incentives, and reputation to local governments that initiated eco-industrial park and low-carbon urban development projects in the same line with the SEPA.

Nonetheless, local governments have sought for local economic growth at the cost of environment and society. This is in part due to the fact that the central communist party still places priority on economic performance in evaluating local government leaders. This has pushed local governments to compulsory acquisition of land to develop industrial plants and real estate with a meager amount of compensation, excusing that land is under state ownership. While the central government mandates an environmental assessment prior to the development, developers do it as an add-on without significant revision in view of the environment and only after the approval of the project. This has in part increased environmental pollution and the number of landless farmers.

By contrast, the Malaysian government has not recognized environmental deterioration as a bottleneck on economic growth. It has framed conversion of rainforest for palm oil plantation as an inevitable price to be rich. It did not regard it necessary to implement stringent regulations and strict enforcement of Malaysian plantation companies since haze pollution suffered by Malaysia and Singapore mainly originates on Sumatra and Kalimantan Island in Indonesia and the existence of a variety of polluters, ranging from local smallholders, transmigrated people toward large-scale Indonesian as well as Malaysian companies, has blurred the responsibility.

In the meanwhile, international NGO and European countries showed their concern, requesting sustainable palm oil production that takes environmental impacts into consideration. In response, the WWF and Malaysian Palm Oil Association (MPOA) established the Roundtable on Sustainable Palm Oil (RSPO), setting 8 principles and 43 standards for sustainable production and granted an RSPO-certified label for palm oil that satisfies these standards. However, MPOA has interfered with implementation of measures to promote sustainable production, claiming that the additional cost to obtain RSPO certificate cannot be recovered by export sales with the premium price. The MPOA goes further to wonder if it should withdraw from RSPO. In the meantime, the Malaysian government and research institutions are advancing research and commercialization of zero-emission palm oil farms and factories to reduce environmental impact of palm oil production in a way that companies can still obtain profit.

5.5 Summary and Prospects

The above experiences of environmental policies in East Asia demonstrate how East Asian country governments have reconciled environmental policies with economic growth. An environmental policy poses political risks to governmental party as it has effects on economic activities and income distribution, including their vested interests. The government does not recognize environmental challenges as problems to be solved and has no incentive to bring in policies that may risk its legitimacy and political and economical support unless it faces fierce protests and pressures against environmental challenges. It was not until the government recognizes that environmental policies will enhance its legitimacy and political support, that is, benefits of implementing environmental policies exceed the cost of leaving them untouched, that it begins to implement environmental policies. However, it tends to frame the challenges that can protect or even increase its political and economic benefits and to implement policy instruments. This leads to the choice of policy instruments that do not properly address the underlying causes and/or have negative implications to macroeconomic stability in expanding government expenditure too much or socially vulnerable people. As long as the government avoids proper framing of underlying causes, implemented environmental policies do not effectively address the negative impacts of economic growth on environment or society, which may become a bottleneck on economic growth, leading to division in the society and loss of legitimacy of the party in power.

In order to avoid such a situation, it is indispensable for the government to frame environmental challenges and to define their underlying causes properly. Then all the government ministries and the political leaders should share the proper framing and definition so that they will take the environment into account in implementing their sectoral policies, in other words, implement preventive measures and convince people and firm to integrate the environment into their activities. Environmental policies should be evolved to go beyond the regulation of individual plant and project, specifically, policies that advance green supply chain management; long-term strategies and plans with ambitious emission reduction target, in consideration for environmental impacts by different types of emission sources; and to allocation of national budget and grant in consideration for their environmental impacts (Mori 2013).

Implementing these preventive and integrative policies would claim economic and political cost in a short term. However, commitment to a long-term ambitious target and gradual implementation with flexible measures enables the government to avoid international and domestic pressures for rapid implementation while convincing people and firm to place it into their core activities. This holds especially true if all the political parties agree on the long-term ambitious targets, or the policy-making process is changed to institutionalize consideration of the policy impacts on the environment. Such kind of environmental policies is expected to reduce political, economic, and social costs to the government, making it brave enough to frame environmental challenges properly, to define and address the underlying causes properly, and to plan to implement environmental policies for the benefit of all in East Asia.

References

Kitagawa H (author and editor) (2011) *Current situation and issues of executing policies of environmental law in China*. Koyo Shobo, Kyoto

Mori A (ed) (2012a) *Democratization, decentralization and environmental governance in Asia*. Kyoto University Press, Kyoto

Mori A (author and editor) (2012b) *Environmental policies in East Asia*. Showado Co. Ltd., Kyoto (in Japanese)

Mori A (author and editor) (2013) Environmental Policy Integration – reform in policy making process in Japan and Europe, and its implementation in transportation sector. Minerva Shobo Co. Ltd., Kyoto (in Japanese)

Mori A, Pengchai P, Sasaki S (2010) "Thailand: retarded environmental policies in political chaos", Japan Environmental Council "State of Environment in Asia" Editing committee ed. State of Environment in Asia 2011/12. Toyo Keizai Inc., Tokyo, pp 241–249 (in Japanese)

Part II
Environmental Technology

Chapter 6
Analysis Methods for Air, Water, and Soil

Yuko Ishibashi

Abstract Composition analysis of air, water, and soil often involves the use of the same type of equipment, though the sampling and pretreatment methods differ. In this chapter, composition analysis is separated into organic analysis and inorganic analysis, including pretreatment methods, with brief explanations of typical equipment. Gas chromatography [GC (FID), GC (ECD)], mass spectrometry [GC/MS, LC/MS], and high-performance liquid chromatography [HPLC] are introduced in the discussion of organic analysis and ion chromatography [IC], atomic absorption spectrometry [AAS], and inductively coupled plasma [ICP-AES/MS] are introduced in the discussion of inorganic analysis.

Keywords Analyzer • Composition analysis • Organic analysis • Inorganic analysis

6.1 Introduction to Analysis

Environmental analysis does not simply consist of determining values with equipment. For example, suppose river water is being analyzed. When sampling river water, a bucket is lowered from a bridge into the river with a length of rope. A large quantity of moss that was present on the handrail of the bridge may have fallen into the bucket when it was lifted. You then bring up the river water containing moss. However accurately you measure it, can it be called a true analysis of the river water? Moreover, any impurities present in the bucket itself also need to be considered. Some items undergo change depending on the preservation condition, even if the sampling is carried out flawlessly. As is seen here, sampling and preservation methods of specimens may often have a direct impact on the analytical value. Judging from these instances, fieldwork should be considered as merely the beginning of analysis. It is suggested that the work procedure (e.g., sampling process) should be confirmed by recording it (on paper) when analysis begins for the first time (details will be provided in another chapter). Accordingly, this chapter describes only analysis methods. Not everything can be described due to space

Y. Ishibashi (✉)
Fukuoka Institute of Health and Environmental Sciences, Fukuoka, Japan
e-mail: y-ishibashi@fihes.pref.fukuoka.jp

T. Shimaoka et al. (eds.), *Basic Studies in Environmental Knowledge, Technology, Evaluation, and Strategy*, DOI 10.1007/978-4-431-55819-4_6

limitations. Therefore, the composition analysis methods introduced here are those involving larger pieces of equipment.

6.2 Organic Analysis

Some organic substances in the environment are highly volatile or have high water solubility due to high polarity. Others are insoluble in water and have a high boiling point. The equipment is selected according to the nature of the substance to be analyzed. Pretreatment methods vary, depending on the nature of the air, water, or soil. Here, brief explanations of some of them are given, including pretreatment methods.

6.2.1 Gas Chromatography with Flame Ionization Detector (GC (FID))

This method is used to analyze trimethylamine, a malodorous substance in the air. After collecting a sample of a fixed volume, a certain amount is poured into equipment that decomposes and concentrates the specimen, which is then collected in a specimen concentration pipe and deeply chilled (with liquid nitrogen or some similar very cold substance). The specimen is then fed into GC (FID) for measuring.

For water, this device is also used to determine the composition ratio of an isomer of nonylphenol, an item used in Environmental Quality Standards, concerned with preservation of aquatic life.

GC (FID) equipment is shown in Fig. 6.1a although it is an older type. Figure 6.1b shows a chromatogram of nonylphenol. There are a number of other possible GC detectors including an electron capture detector (ECD), a thermal conductivity detector (TCD), a flame thermionic detector (FTD), flame photometric detector (FPD), and a helium ionization detector (HID).

Fig. 6.1a GC (FID)

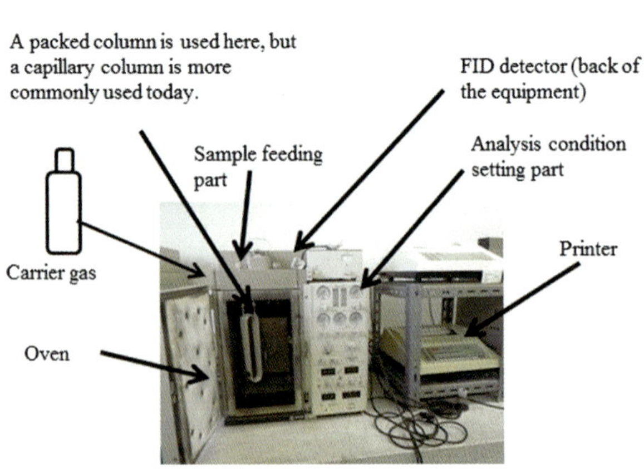

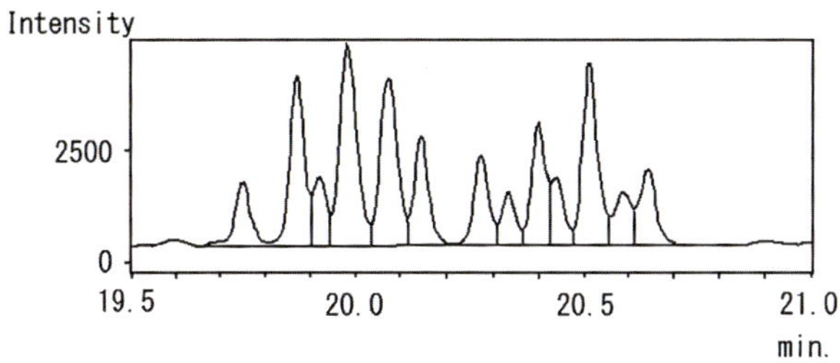

Fig. 6.1b Chromatogram of nonylphenol measured with GC (FID)

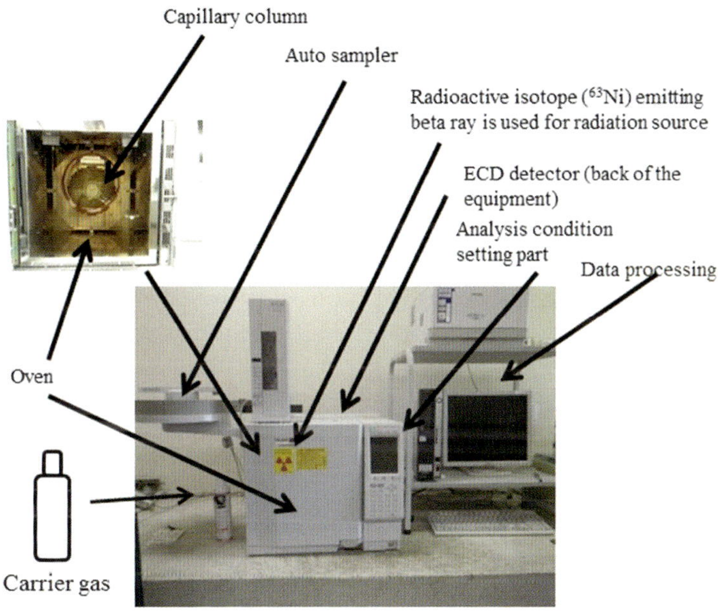

Fig. 6.2a GC (ECD)

6.2.2 Gas Chromatography with Electron Capture Detector (GC (ECD))

The Water Pollution Control Law stipulates that PCB must be analyzed using the GC (ECD). A sample is hexane extracted and dehydrated for concentration. It is then put through a chromatographic tube filled with silica gel and concentrated after being cleaned up. Then, its volume is fixed. This is fed into the GC (ECD) to be measured. Figure 6.2a shows GC (ECD) equipment. Figure 6.2b shows a chromatogram of PCB.

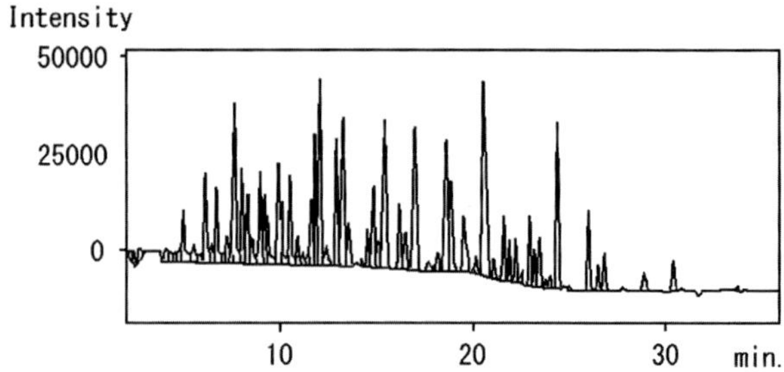

Fig. 6.2b Chromatogram of PCB measured with GC (ECD)

6.2.3 Gas Chromatography Mass Spectrometry (GC/MS)

GC/MS is one the most frequently used items of equipment for analysis of environmental organic substances. There are various types of sample feeding parts for different substances.

In the first type, samples gained by solid-phase extraction or liquid-liquid extraction are directly fed in (microsyringe or autosampler). Simazine, which is an item in Environmental Quality Standards in the Water Pollution Control Law, is analyzed using this type of equipment. In the second type, a headspace (HS) sampler is connected to the sample feeding part. Highly volatile organic matter is gasified by adding salt to a water sample and heating it. It is then fed into the GC/MS. Benzene, which is an item in Environmental Quality Standards in the Water Pollution Control Law, is analyzed using this type of equipment. In the third type, a purge-and-trap (PT) device is connected to the sample feeding part. Highly volatile organic matter is transformed to its vapor phase by running gas in the water sample and is adsorbed on the adsorbent with a trap pipe. It is then rapidly heated, and desorbed gas is fed into the GC/MS. Epichlorohydrin, which is a monitoring-required water quality item in the Water Pollution Control Law, is analyzed using this type of equipment. For the fourth type, a canister can be connected to the sample feeding part. For toxic air pollutants listed in the Air Pollution Control Law (e.g., tetrachloroethylene), samples are collected with canisters and fed into the GC/MS.

Figure 6.3a shows the GC/MS equipment. Figure 6.3b shows a chromatogram of simazine. These days, GC/MS/MS equipment is on the market, enabling identification of compound substances that have been difficult previously. This development has widened the application range.

6.2.4 High-Performance Liquid Chromatography (HPLC)

Thiuram, which is an item in the Environmental Quality Standards in the Water Pollution Control Law, is analyzed using this type of equipment. It is converted and dissolved into acetonitrile solvent through solvent extraction or solid-phase

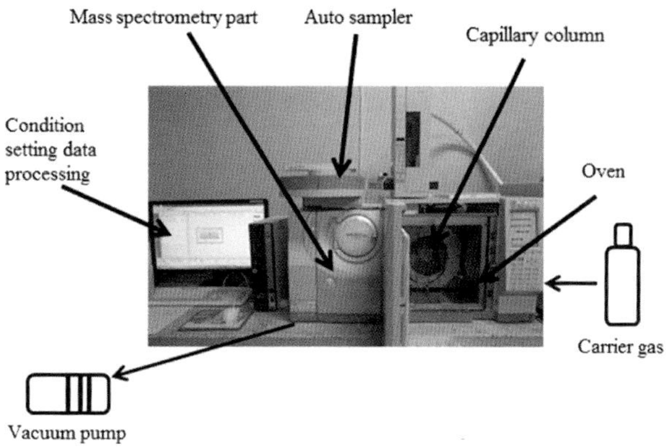

Fig. 6.3a GC/MS

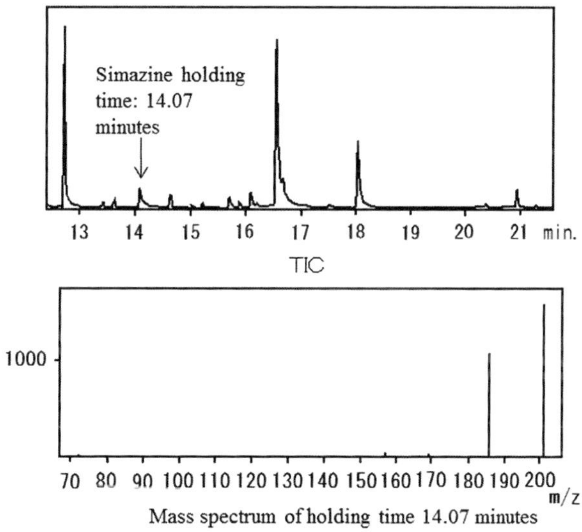

Fig. 6.3b Chromatogram of simazine measured with GC/MS

extraction and measured in HPLC. An ultraviolet absorption detector (272 nm) is used. Oxine-copper, which is a monitoring-required water quality item in the Water Pollution Control Law, and anionic surfactant, which is a water quality standard item in the Water Supply Act, are measured with the HPLC.

One other type of HPLC detector is a fluorescent detector (see Fig. 6.4a). Figure 6.4b shows a chromatogram of thiuram.

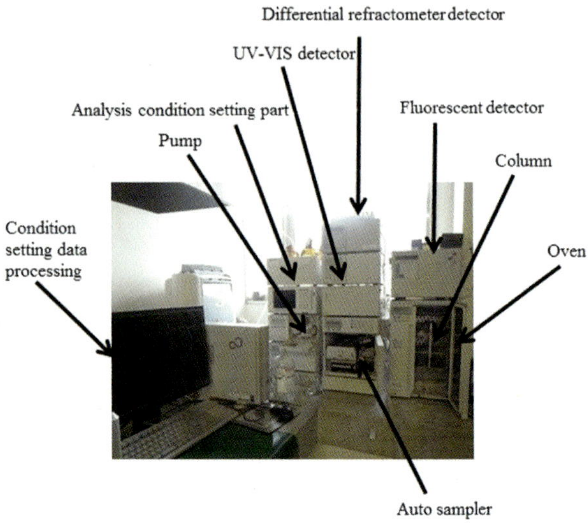

Fig. 6.4a HPLC

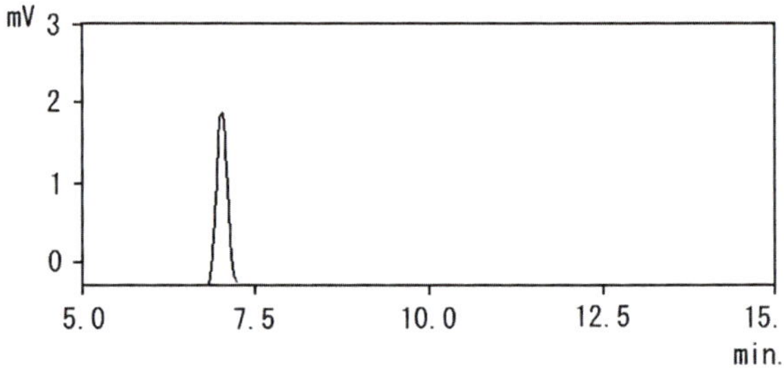

Fig. 6.4b Chromatogram of thiuram measured with HPLC

6.2.5 Liquid Chromatography-Tandem Mass Spectrometry (LC/MS/MS)

Linear alkylbenzene sulfonate (LAS), which is a water quality standard item concerning preservation of aquatic life, in the Water Pollution Control Law, is analyzed using this type of equipment. It is measured in LC/MS/MS after solid-phase extraction. This equipment is also highly effective for analyzing types of organic matter with high polarity.

No database of LC/MS/MS has been made since the gained mass spectrum varies between the types of equipment. Moreover, measurement varies depending on the measuring condition. As a result, it is not currently appropriate for identifying unknown samples. Figure 6.5a shows the LC/MS/MS equipment, and Fig. 6.5b shows a chromatogram of LAS.

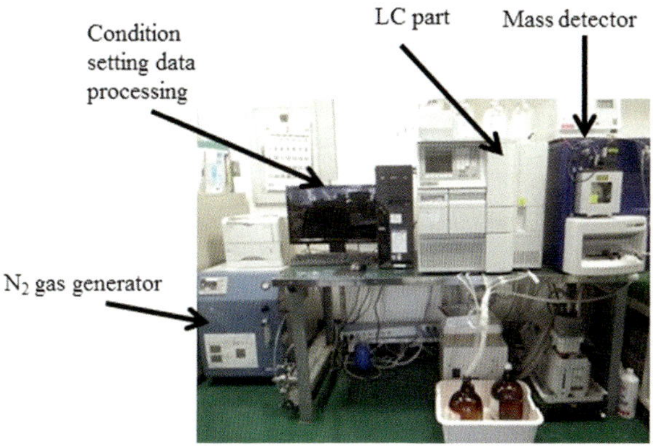

Fig. 6.5a LC/MS/MS

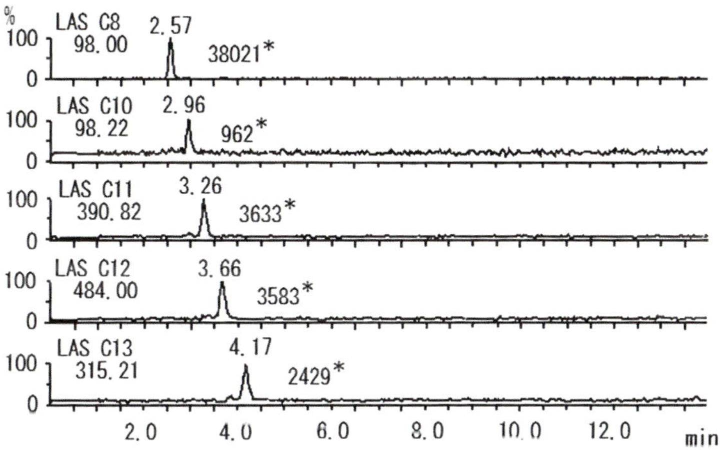

Fig. 6.5b Chromatogram of LAS measured with LC/MS/MS

6.3 Inorganic Analysis

Pretreatment of inorganic matter in the environment is relatively easy, if involves measuring total or soluble (e.g., filtered through a membrane filter) substances. However, morphological analysis is often difficult. This chapter deals with total or soluble substance analysis, including pretreatment methods, and parts of it are briefly explained.

6.3.1 Ion Chromatography (IC)

This type of equipment is often used for analyzing acid rain monitoring surveys and microscopic particles (e.g., PM2.5) for air and chloride ion for water assessed according to the Water Supply Act. This equipment is often used for inorganic ionic analysis. Filtered sample solution is fed into the IC for measurement. In general, a conductivity detector is used, but an ultraviolet (UV) detector may be used for nitrate ions. In addition, an electrochemical detector (ECD) is available. Figure 6.6a shows the IC equipment. Figure 6.6b shows an anionic chromatogram.

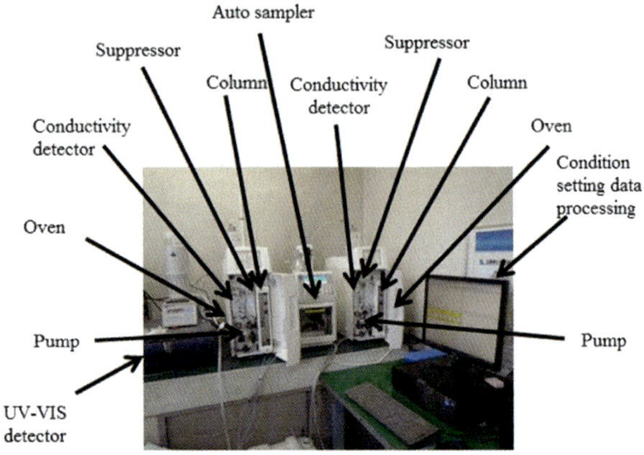

Fig. 6.6a IC

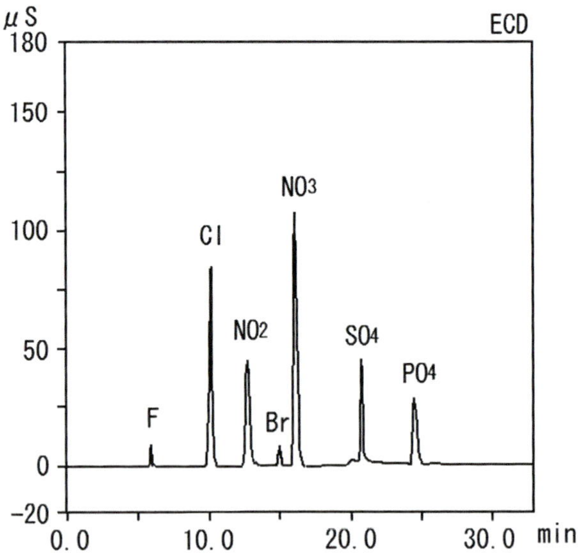

Fig. 6.6b Anionic chromatogram measured with IC

6.3.2 Atomic Absorption Spectrometry (AAS)

This is used to analyze metals that are contained in samples. There are two types of AAS: the flame type and the electric-heated type (graphite furnace, also called flameless). They have different ways of atomizing metals contained in samples. The electric-heated AAS has higher sensitivity when compared with the flame AAS, but the volume of sample that can be fed is minimal, which may result in substantial variability. This is why the flame AAS is often preferred.

Organic solvent can be fed directly into the AAS. Therefore, metals in samples are concentrated or cleared of interfering substances after complexation and solvent extraction, for AAS measurement.

6.3.3 High Frequency Inductively Coupled Plasma Atomic Emission Spectrometry/Mass Spectrometry (ICP-AES/MS)

This device, which is introduced in the next section, is another item of equipment for analyzing metals. This has numerous superior features over AAS, starting with a wider dynamic range. This enables multi-element analysis or allows more elements to be assessed with high sensitivity. However, the AAS has lower cost of operation. Therefore, it is still often used in routine work. Figure 6.7 shows the flame AAS equipment.

In addition, the AAS can measure arsenic, selenium, and antimony, when hydride generation apparatus is used.

Mercury is highly volatile, so it has to be measured separately. A reduced vapor AAS is on the market exclusively for analyzing mercury. Figure 6.8a shows the reduced vapor AAS equipment. Figure 6.8b shows a measurement chart of mercury.

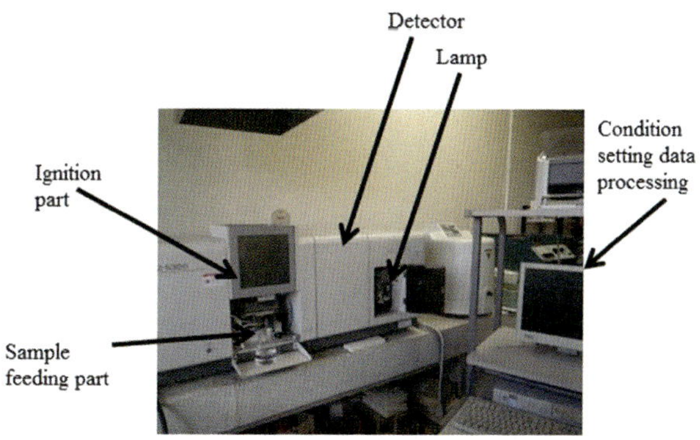

Fig. 6.7 Flame AAS

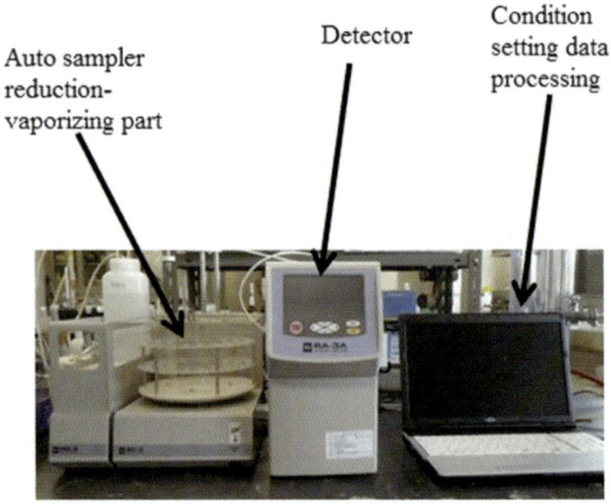

Fig. 6.8a Reduction vapor AAS

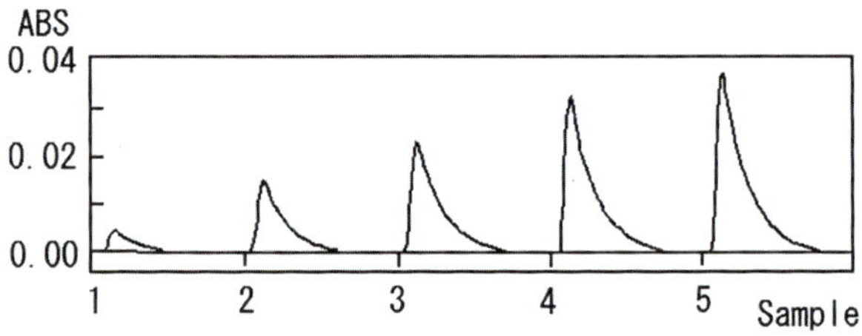

Fig. 6.8b Chart of mercury measured with reduced vapor AAS. Results from five standard mercury samples of varied concentration

6.3.4 High Frequency Inductively Coupled Plasma Atomic Emission Spectrometry/Mass Spectrometry (ICP-AES/MS)

The ICP-AES/MS is also used to analyze metals. It is currently the most commonly used analysis equipment for metals. The ICP uses a light source generated by a high-temperature toroidal plasma flame, produced by heating ionized argon gas with induced current in a coil with high-frequency power output. Metals can be efficiently analyzed by spraying a sample solution. IPC-AES and ICP-MS are explained separately in the following section.

ICP-AES consists of two types: sequential and multichannel. The sequential type is good for analysis with high sensitivity, but this one only analyzes one element at a time. The sensitivity of the multichannel type is inferior to that of the sequential type; however, it does allow multi-elemental analysis. The sequential type is often used in the field of raw materials, while the multichannel type is used in the environmental field.

ICP-MS allows multi-elemental analysis, and it has the highest sensitivity of all metal analyzers. Cadmium, which is an item in the Environmental Quality

Standards in the Water Pollution Control Law, and zinc, which is an item in Environmental Quality Standards concerning the preservation of aquatic life, can easily be analyzed using ICP-MS. This type is used for component analysis of PM2.5 and for rare metals, which is attracting particular attention today. Figure 6.9a shows the ICP-MS equipment. Figure 6.9b shows a spectrum of zinc.

The pretreatment for analyzing metals in PM2.5 is to decompose them using pressure and heat, in a decomposing apparatus pressurized vessel or hot decomposition on a hot plate after adding acid to a filter which has trapped PM2.5 (cut into adequate sizes). This pretreatment method can be used for total decomposition of soil and substratum. However, total decomposition based on the Soil Contamination Countermeasure Act is a dissolution test using hydrochloric acid.

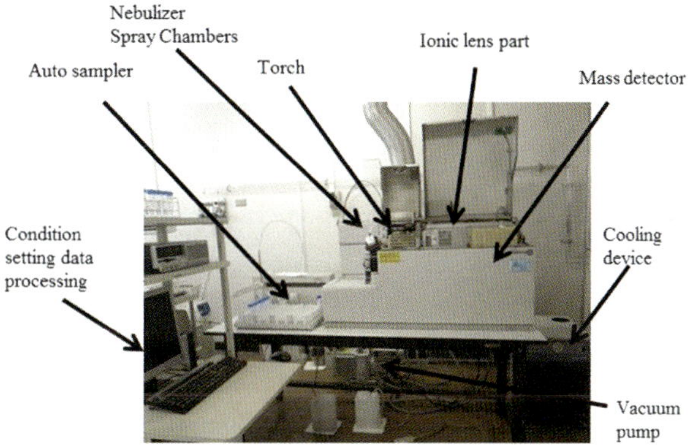

Fig. 6.9a ICP-MS

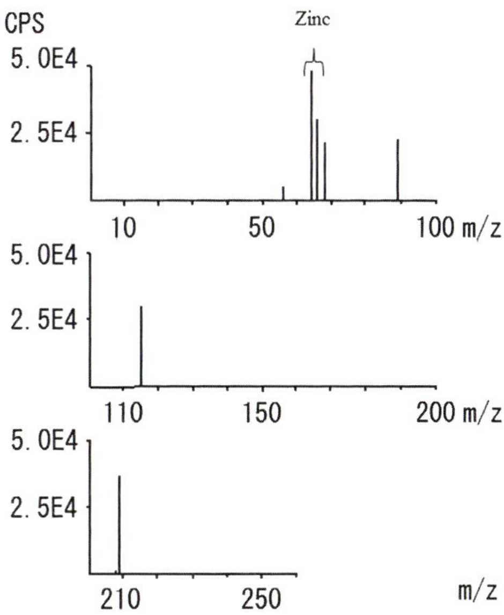

Fig. 6.9b Spectrum of zinc measured with ICP-MS. Internal standard substances: yttrium, indium, bismuth

Hydrochloric acid cannot be fed into the ICP-AES/MS without treatment. Therefore, hydrochloric acid must be completely evaporated to convert it to nitric acid for feeding into equipment. Hydrochloric acid can be used for analysis with the AAS.

6.4 Ending Analysis

The final stage after gaining an analysis result is the report, and I want you to review the result again. It can be said that for any analysis method, there is a possibility of some form of interference. Recent equipment is all controlled by computer, and it is often the case that only the result values are printed out. Do not accept the values without verification. You should never neglect the confirmation work (e.g., for each peak). Detailed principles are not described in this chapter. However, I want you to understand everything, from the beginning of the analysis to the details of analysis methods including pretreatment methods, so that you can present reports with confidence, on the results you obtain.

Chapter 7
Environmental Quality Analysis

Yuka Watanabe

Abstract Environmental quality assessments require the determination of concentrations of various substances. This chapter outlines key points for environmental quality analysis. First, environmental standard values (environmental standards and regulation standards) and analytical methods (official methods) are explained. Next, preparation methods for analytical samples are discussed with respect to sampling methods and sample preparation. Samples must be suitable for analysis after preparation. Pretreatment of samples for analysis is discussed in Sect. 6. Analytical values are obtained by the analysis of pretreated samples, and the reliability of the analytical values is described in Sect. 7. Finally, a leaching test, used for waste analysis, is discussed in terms of judgment criteria for landfill waste, environmental quality standards for soil pollution, soil leaching and content standards, and radioactive substance standards.

Keywords Environmental standards • Official method • Sampling • Pretreatment • Reliability of analytical values • Leaching test

7.1 Introduction

Environmental quality analysis is conducted for various reasons, and thus types of environmental samples vary greatly, including water, air, soil, and waste samples. Environmental quality analysis also covers a wide range of processes, including the preparation of a sampling plan and its execution, transportation and storing of samples, chemical analysis (pretreatment and measurement), and statistical treatment of data and interpretation. Among these processes, chemical analysis requires a complicated analytical method for each sample, due to the fact that there are diverse chemical substances with different concentrations and properties in samples, and some substances may interfere with analysis of other substances.

The first half of this chapter on environmental quality analysis introduces environmental standards, official methods, data reliability, and pretreatment of samples, and the second half of this chapter discusses a case involving waste

Y. Watanabe (✉)
Research Institute for East Asia Environments, Kyushu University, Fukuoka, Japan
e-mail: watanabe@rieae.kyushu-u.ac.jp

T. Shimaoka et al. (eds.), *Basic Studies in Environmental Knowledge, Technology, Evaluation, and Strategy*, DOI 10.1007/978-4-431-55819-4_7

analysis. There are various kinds of waste analysis, including hazard judgment tests of wastes, content tests for recycling, and trace analysis. This chapter describes a leaching test, a type of hazard judgment test.

7.2 Environmental Standards

With the aim of protecting human health and conserving the environment, environmental standards (ESs) are used to establish benchmarks for air, water, soil, and noise quality. Benchmarks are then used to create policies that will maintain those ESs. ESs serve as a “desired goal,” i.e., a target used by governments to establish environmental policies. They are not representative of the minimum level needed to maintain human health and other conditions; rather, they are a set of targets that are meant to be actively maintained. For areas in which pollution is not currently taking place, ESs should be established in order to maintain the status quo and therefore avoid any decline from the current state (Website of the Ministry of the Environment). ESs are defined in Japan’s Basic Environment Law (Basic Environment Law Section 3 “Environmental Quality Standard”). In the Basic Environment Law, concrete values of standards are established for air (Environmental Quality Standards for Air), noise (Environmental Quality Standards for Noise, Environmental Quality Standards for Aircraft Noise, and Environmental Quality Standards for Shinkansen Super-express Railway Noise), water (Environmental Quality Standards for Water and Environmental Quality Standards for Groundwater), and soil (Environmental Quality Standards for Soil) quality. Concrete values of standards are also defined for dioxins released through air pollution, water contamination (including contamination of the seafloor), and soil pollution under the Act on Special Measures against Dioxins (Website of the Ministry of the Environment).

7.3 Regulation Standards

Regulation standards (RSs) are a set of standards to regulate behavior that causes pollution as defined by laws or ordinances. Facilities such as factories are required to comply with these standards. Terms vary depending on the laws, i.e., the RSs are called “emission standards” in the Air Pollution Control Act, “effluent standards” in the Water Quality Pollution Control Act, and “regulatory standards” in the Noise Regulation Law, the Vibration Regulation Law, and the Offensive Odor Control Law. RSs are generally imposed to maintain ESs.

7.3.1 Stringent Standards

National governments establish RSs for the emission of dust, soot, and toxic substances into the air and various effluents into the water. Prefectural governments are allowed to establish more stringent standards than national standards through ordinances based on their judgment of natural and social conditions. Such standards are called stringent standards (SSs).

7.3.2 Total Amount Regulation

Total amount regulation (TAR) is a method used to assign the allowable emission of pollutants and contaminants from factories, thereby limiting the total level of pollutants and contaminants in a certain area to an allowable level in terms of environmental conservation. The allowable level is determined by multiplying the gas emission volume by the concentration of the pollutant for air pollution and by multiplying the drainage volume by the concentration of the contaminant for water contamination. TAR is enforced in specific areas for sulfur oxides and nitrogen oxides in the air and for chemical oxygen demand (COD), nitrogen, and phosphorus in the water.

7.4 Official Method

An "official method" is important for environmental quality analysis and frequently used by governments for performing environmental monitoring. For example, standard values and measurement methods for dioxins are defined by ESs related to air pollution, water contamination (including contamination of the seafloor), and soil pollution by dioxins (Website of the Ministry of the Environment). Similarly, water quality is based upon Japanese Industrial Standards (JIS) K 0312 (2008). A "measurement manual" is also established for an "official method," defining detailed measurement methods. The use of an "official method" is necessary for analytical values to become official values, and a certified environmental measurer must issue a measurement certification. The measurement certification also verifies the ability of the measurer.

7.5 Sample Preparation

7.5.1 *Sampling (Collecting Samples)*

When sampling is performed for analysis, it is important that the sample is collected without bias and is representative of the whole population of interest. Generally, the whole population of environmental samples is heterogeneous, and therefore the sampling technique requires careful attention. Increasing the number of samples will improve the estimation of the representative value of the whole population, but the cost of sampling and time constraints may make it difficult to collect more samples.

Types of sampling methods include random sampling and systematic sampling. In the former method, samples are chosen randomly and entirely by chance; in the latter method, samples are collected at a constant interval of time and space. There are also other sampling methods, such as two-stage sampling, and an appropriate sampling method should be chosen based upon the state of the samples.

7.5.2 *Sample Preparation*

A lot, or batch, is a set of materials subjected to analysis that come from the same source, with nearly uniform properties and constituents. The amount of sample collected from the lot by a sampler in a single action is called an increment. For example, a solid sample (e.g., soil) is tested using a shovel designed for increment sampling, and each collected increment is pulverized separately. All increments are combined into a bulk sample, or gross sample, in order to obtain a representative sample with average composition for the whole analytical population and then subjected to "sample division" using either an increment sample division method, a method with a riffle sampler, or a sample division apparatus to obtain laboratory samples. Sample division is a process by which a single sample is divided into smaller groups of samples with the same physical properties or chemical constituents. Divided samples are then used as analytical samples. Figure 7.1 illustrates the sampling and sample preparation procedure for soil. The details for sampling and sample procedure for wastes are shown in JIS K 0060 (1992).

Once the samples from the bulk sample are used up, it is very difficult to obtain identical environmental samples with the same properties at a later time. The common practice for analysis is, therefore, to refrain from using up the samples during the preparation procedure and set aside a portion of the samples.

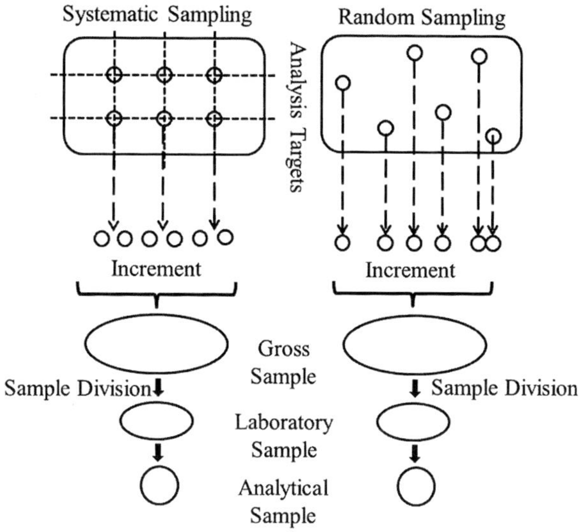

Fig. 7.1 Soil sampling and sample preparation procedures

7.6 Pretreatment

Increments collected from analytical populations by sampling are transformed into analytical samples through a series of techniques. Raw samples are subjected to analysis if they can be used directly for instrumental analysis or titration. However, samples usually require pretreatments such as decomposition, dissolution, extraction and concentration of the target constituent, and the removal of inhibiting constituents. Main techniques of pretreatment are detailed below.

7.6.1 Decomposition and Dissolution

If an analytical sample is converted into a solution, the sample is processed by wet decomposition, generally using an acid (e.g., hydrochloric acid, nitric acid, sulfuric acid). Stronger acids often used in wet decompositions include perchloric acid and hydrofluoric acid. Perchloric acid must not be used alone on organic compounds because the reaction is explosive and very dangerous. Hydrofluoric acid is also hazardous. If hydrofluoric acid adheres to and then penetrates the skin, it causes a chemical burn that is difficult to heal. Therefore, gloves are mandatory when using hydrofluoric acid. Inhaling the vapor of hydrofluoric acid is even more hazardous. When decomposing a sample, the reagent must not interfere with the following analytical techniques. In the past, heating during decomposition was conducted using an electric furnace or a hot plate, but have been replaced by microwave extraction in recent years. In this method, samples are sealed in a Teflon pressure decomposition container and decomposed by irradiating microwaves. Therefore, this method prevents the volatilization of acids and can decompose the sample with a small amount of acid in a short period of time. Another decomposition process is the dry ash method. However, this process is an open system that carries the risk of pollution and scattering of samples and acid. Therefore, a closed system has been favored for pretreatment in recent years.

7.6.2 Extraction from Solids

The pretreatment method in which an analytical target substance (analyte) is taken out of a solid sample is called extraction. A simple method involves combining the sample with a solvent, pulverizing with a homogenizer to uniformity, mixing with a shaker, and then filtering. Soil samples are typically treated using a Soxhlet extractor. When using a Soxhlet extractor, a sample is placed on a cylindrical filter paper and extracted with an organic solvent through a cycle of repeated evaporation and condensation. Other methods of extraction from solids include supercritical fluid extraction.

7.6.3 Liquid-Liquid Extraction

Transferring analytes or interfering substances in a sample solution to another solvent is called liquid-liquid extraction. Solvents immiscible with each other are added into a separable funnel with the sample, and the separable funnel is then shaken until the two layers form an interface and reach the distribution equilibrium. The solution is then left standing, until the interface separates completely, and the layer of interest is taken out. Generally, organic compounds are transferred from an aqueous solution to an organic solvent. In the case of metal analysis, chelating agents may be used to extract the metal as the metal chelates.

7.6.4 Solid Phase Extraction

Solid phase extraction is a separation method in which a gas or liquid sample is passed through a solid phase to trap chemical constituents. The use of solid phase extraction is expanding for use in other applications, such as the removal of coexisting substances by use of different adsorbents (solid phase) or solvents, in addition to the separation (extraction) of analytes. Solid phase extraction methods include the batch method, column method, and membrane method, although the cartridge method is most commonly used. Various solid phases have been used in studies, which include a normal-phase type, an ion-exchange type, a chelate type, and a reverse-phase type. The cartridge is first conditioned by washing with the solvent to wet the surface, and then the sample solution is added to the cartridge. After the sample has been loaded into the cartridge, impurities are removed through washing with a nonpolar solvent. Finally, the analytes are eluted with a polar solvent.

7.6.5 Concentration

Concentration is a process by which some amount of solvent is removed from a solution to increase the concentration of the target solute in the solution, used in cases where only trace amounts of the analyte exist in the solution sample. Concentration is necessary especially when the concentration is lower than the minimum detectable concentration (limit of detection) of the analytical instrument. A common practice is to use a rotary evaporator, in which the sample solution is heated, rotated, and depressurized, and the solvent is evaporated and recovered by cooling.

For a small amount of solution sample, nitrogen gas is blown in to evaporate the solvent to concentrate the solution. A Kuderna-Danish (KD) concentrator and a centrifugal evaporator are also used. However, these concentration methods have a drawback, as they also concentrate coexisting substances along with the analytes. It is possible to separate coexisting substances by a liquid-liquid extraction method or solid phase extraction method, in addition to concentration.

7.6.6 Other Pretreatment Methods

Distillation Distillation is a fundamental pretreatment method that has been used since the early days of chemical measurements. Types of distillation include steam distillation and reductive distillation. Distillation can remove impurities that may interfere with analysis. Distillation separates component substances from liquid mixtures through vaporization and condensation based on different volatilities, and thus, highly pure analytes can be obtained.

Headspace Method and Purge-and-Trap Method The headspace method and purge-and-trap pretreatment methods utilize the process of gasification. In these methods, the gas phase is taken out and analyzed using gas chromatography or other instruments. The headspace method gasifies the analytes when the sample liquid is sealed in a container. The sample rests until it reaches a constant concentration based on its partition coefficient. The sample is then taken out by a gas phase syringe. This method is widely practiced due to its ease of use.

Gel Permeation Chromatography (GPC) Gel permeation chromatography (GPC) is a type of size exclusion chromatography in which molecules are sifted according to their size. The method is able to separate analytes from sample-derived fat constituents.

Other methods of pretreatment include filtration, dialysis, precipitation, and recrystallization.

7.7 Reliability of Analytical Values

7.7.1 Uncertainty

Conventionally, an error has been defined as the "difference between the measured value and true value"; however, a "true value" is a conceptual value that cannot actually be obtained. "Uncertainty" is a term proposed by several international organizations, including the International Organization for Standardization (ISO), as a unified expression of the reliability of measurement results, with the idea that a range of measured quantity is obtained from the dispersion of the data using known measurement results.

With the proposal of the concept of uncertainty, the definition of "reliability" in JIS K 0211 was also updated as the expression of the concrete range of the measured quantity, from "the degree of precision or correctness expected" (1987 version) to "tempostable property or degree to maintain specified functions and performance of instruments, methods, and their elements within the boundary of a given set of conditions" (2013) (JIS K 0211 2013).

7.7.2 Technical Terms Related to Reliability

The degree of dispersion of measured values is called "precision," while the degree of deviance from the true value is called "trueness." The concept that combines precision and trueness is called "accuracy." The quantitative expression of accuracy is "uncertainty." While the degree of deviance of measured values from the true value was referred to as "correctness" in the past, it is referred to as "trueness" today in order to avoid confusion between "correctness" and "accuracy." Precision is further divided into repeatability and reproducibility. Repeatability is the degree of consistency in values, when the same analyst continuously repeats measurements using the same method on the same day. Reproducibility is the degree of consistency in values, when the date and time, analyst, and/or methods are different. Reproducibility yields greater dispersion than repeatability. However, different technical terms may be used in place of these technical terms in some fields, and verification is necessary in each field (this section uses the terminology from the field of chemical measurements).

7.8 Leaching Test Method

Analyses of wastes are roughly classified into leaching tests and content tests. Analyses related to landfill disposal are subjected to leaching tests, which are at the center of waste analysis. A leaching test is the tool used to assess the environmental impact of wastes when the wastes are buried in landfill or recycled. The leaching test method varies by country; this section introduces the official method in Japan.

7.8.1 Judgment Criteria for Landfill

Announcement No. 13 by the Ministry of the Environment ("Test Method for Metals in Industrial Wastes," partially revised in 2013) outlines the method of judging the hazard of toxic substances leached from wastes buried in a landfill. The leaching test defined in the announcement is as follows.

A waste sample is prepared by sifting to obtain particles with diameters of 0.5–5 mm. Pure water (A3 or A4 in JIS K 0557) (JIS K 0557 1998) is used as the solvent. The sample and pure water are mixed at a liquid-solid ratio (L/S, the ratio of 10 g of solid to 1 mL of liquid) of 10. The mixture is subjected to horizontal shaking for 6 h. After shaking, the resulting mixture is subjected to centrifugal separation at a 3000 gravitational acceleration for 20 min. Filtration is performed with a 1 μm membrane filter, and the filtrate is finally submitted for analysis.

L/S is set at 10 with the idea that a criterion is set for the hazard, when waste materials come in contact with ten times the weight of water to the waste (Sakai et al. 1996).

7.8.2 Environmental Quality Standards for Soil Pollution

A leaching test method for measuring environmental quality standards for soil pollution is defined in Announcement No. 46 by the Ministry of the Environment ("On the Environmental Quality Standards for Soil Pollution"). This method is also sometimes used as a judgment method to assess suitability when waste-derived materials are recycled for secondary use for purposes such as construction materials. Samples with particle diameter of 2 mm or smaller are prepared by sieving and then subjected to horizontal shaking under the same conditions as those in Announcement No. 13 by the Ministry of the Environment. The sample is centrifugally separated, and the filtrate is obtained by filtration with a 0.45 μm membrane to be submitted for analysis.

7.8.3 Soil Leaching and Content Standards

A measurement method for soil leaching is defined in "On Determining a Measurement Method for Soil Leaching Surveys (2003 Ministry of the Environment Announcement No. 18)." The quantity of leaching is measured by the appended measurement method from Announcement No. 18 by the Ministry of the Environment, using testing liquid prepared by the method in Announcement No. 46 by the Ministry of the Environment. The appendix of Announcement No. 46 defines measurement methods by the types of specific toxic substances. This measurement method assumes a scenario in which toxic substances in soil pollute the groundwater.

A measurement method for soil content is defined in "On Determining a Measurement Method for Soil Content Surveys (2003 Ministry of the Environment Announcement No. 19)." At least 6 g of the sample as in Announcement No. 46 (sample with 2 mm or less particle diameter prepared with sieve) is used, and the sample is mixed with a solvent, 1 mol/L of hydrochloric acid (unit mL), at 3 % of the weight-volume ratio. The mixture is shaken at 200 rpm for 2 h by a shaker and allowed to stand for 10–30 min. The filtrate is obtained by filtration of the supernatant from the mixture with a membrane filter of 0.45 μm pore diameter to be submitted for analysis. The amount of heavy metals leached into the hydrochloric acid is calculated as the amount per 1 kg of dry sample weight. This measurement method is used to assess the risk for oral intake of metals in the soil.

7.8.4 Radioactive Substance Standards

A great amount of wastes containing radioactive substances were generated due to environmental pollution of radioactive substances released from the Fukushima power plant disaster, caused by the Great East Japan Earthquake on March 11, 2011. For these wastes containing radioactive substances, the leaching test method in JIS K 0058-1 (2005) is used, rather than the methods in Announcement No. 13 by the Ministry of the Environment. This is the method selected to test the leaching property of radiation at waste treatment sites under conditions similar to those in the field. The procedure is detailed in the "Waste Material Guidelines Part 5" of "A Guideline for Measuring Radioactive Concentration" (2011).

Figure 7.2 shows the schematic drawing of the leaching test apparatus. In this method, solvent, ten times the weight of water to the sample, is added to a certain amount of sample, and the resulting mixture is stirred at 200 rpm for 6 h to leach chemical substances (e.g., cesium) and obtain the testing liquid. After stirring for 6 h, the sample is allowed to stand for 10–30 min, and the resulting dispersion is removed from the tank. Removed liquid is centrifugally separated at 3000 rpm for 20 min as needed, and the filtrate of the supernatant is obtained by filtration with a membrane filter of 0.45 μm to be submitted for analysis.

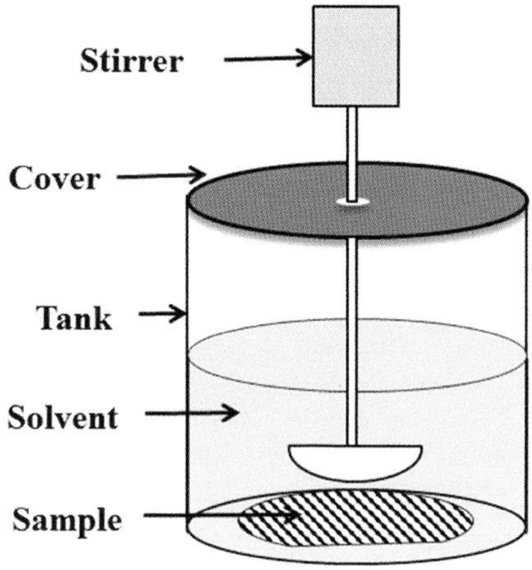

Fig. 7.2 Schematic drawing of the leaching test apparatus

7.9 Conclusion

This chapter provides an overview of some of the key points in environmental analyses. However, types of environmental samples are diverse and include water, air, soil, and waste samples, and analytical methods are equally diverse. This chapter does not go into extensive details, due to space constraints. The reader is referred to the references for further details (Hirai 2006; Ichikuni 1998; Tsunoda 2013; Tsumura 2009). It is very important to consider the quality and quantity of samples for assessing the environment. This chapter is to be used as a reference, and in conjunction with detailed measurement manuals developed by the Japanese Industrial Standards Committee and the Ministry of the Environment, highly reliable analyses may be conducted for environmental impact assessments.

References

Hirai A (editor-in-chief) (2006) The Japan Society for Analytical Chemistry edition: basics of on-site chemical analysis. Ohmsha, Tokyo, Japan

Ichikuni M (ed) (1998) Science and analysis of substances, foundation for the promotion of the Open University of Japan

Japanese Industrial Standards (JIS) K 0312 (2008) Method for determination of tetra-through octachlorodibenzo-p-dioxins, tetra-through octachlorodibenzofurans and dioxin-like polychlorinated biphenyls in industrial water and waste water. Japanese Standards Association, Japan

JIS K 0058-1 (2005) Test methods for chemicals in slags—part 1: leaching test method. Japanese Standards Association, Japan

JIS K 0060 (1992) Sampling method of industrial wastes. Japanese Standards Association, Japan

JIS K 0211 (2013) Technical terms for analytical chemistry (general part). Japanese Standards Association, Japan

JIS K 0557 (1998) Water used for industrial water and wastewater analysis. Japanese Standards Association, Japan

Sakai S, Mizutani S, Takatsuki H (1996) Leaching tests for waste materials. J Jpn Soc Mater Cycles Waste Manag 7(5):383–393

Tsumura Y (2009) Basics and mechanisms of the latest in analytical chemistry, Shuwa System, Japan

Tsunoda K (2013) The Japan society for analytical chemistry edition: environmental analysis (analytical chemistry practice series (applied analysis #6) Kyoritsu Shuppan

Waste materials guidelines part 5, "Guidelines for measuring radioactive concentration" (Dec. 2011, 2nd edition March 2013)

Website of the Ministry of the Environment. ESs related to air pollution, water contamination (including contamination of the sea floor), and soil pollution by dioxins. http://www.env.go.jp/kijun/dioxin.html

Website of the Ministry of the Environment. On the environmental quality standards of Japan. https://www.env.go.jp/kijun/

Chapter 8
Water Purification by Reducing the Pollution Burden

Hiroshi Kumagai and Takahiro Kuba

Abstract This chapter introduces Fukuoka Prefecture as an example of measures taken by an administrative body toward preservation of the aquatic environment. Explanations are given about various standard items concerning water quality. These are the decision-making standards that apply when measures must be taken. In addition, methods to reduce the pollution burden of water, as well as methods to encourage natural purification, are introduced. The water monitoring condition, current condition of water quality, and improvement in water quality in the past 40 years are shown. At the same time, a description is given of problems that will be encountered when proceeding with aquatic habitat preservation from now on.

Keywords Pollution burden • Water pollution control law • Environmental quality standards • Effluent standard • Public bodies of water • Constant monitoring

8.1 Overview of Water Pollution

8.1.1 Reduction of Water Pollution

Water pollution occurs when more pollutants are introduced than a water body is capable of purifying. This may lead to problems such as health hazard, offensive odor, damage to marine and agricultural products, and fatal effects on aquatic life.

In order to prevent water pollution from occurring, measures are taken either to reduce the amount of polluting materials in the water (pollution burden) or to reduce the pollutant effect by increasing the purification capacity of the water body.

Because an enormous amount of cost and labor is involved in purifying the pollutants present in water, it is essential to take measures to prevent pollutants

H. Kumagai (✉)
Environmental Preservation Division, Department of Environmental Affairs,
Fukuoka Prefectural Government, Fukuoka, Japan
e-mail: kumagai@fihes.pref.fukuoka.jp

T. Kuba
Department of Urban and Environmental Engineering, Graduate School of Engineering,
Kyushu University, Fukuoka, Japan

T. Shimaoka et al. (eds.), *Basic Studies in Environmental Knowledge, Technology, Evaluation, and Strategy*, DOI 10.1007/978-4-431-55819-4_8

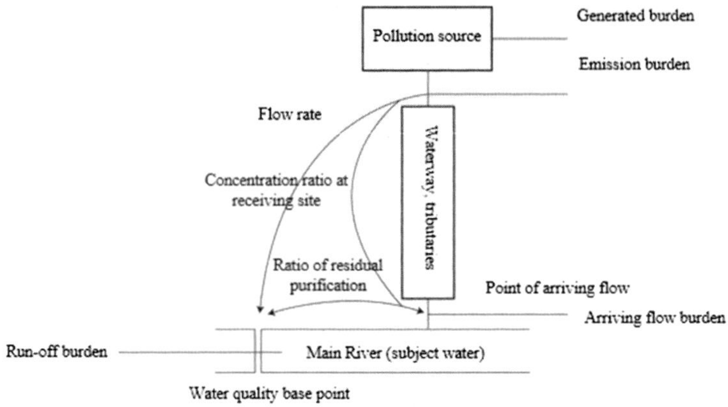

Fig. 8.1 Conceptual diagram of flow burden (Japan Sewage Works Association 2008)

from reaching our water bodies. In fact, administrative water pollution prevention measures are mainly targeted at incoming pollution burden. It is important to ascertain where pollutants are generated and the amounts involved, well before they enter the water.

8.1.2 Pollution Burden

Figure 8.1 (Japan Sewage Works Association 2008) is a schematic concept of the pollution burden. Pollution can be divided into (1) generated and (2) incoming pollution burdens, and different measures are taken to deal with each.

The generated pollution burden is that which is generated in the subject basin area and which then flows into rivers in that area. This method aims to reduce or eliminate emitted substances at their sources. The "Stockholm Convention on Persistent Organic Pollutants" (POPs Convention) requires regulation of persistent organic compounds with a high residual effect, such as PCB and DDT, for member states of the Convention. In Japan, discharge into public bodies of water is restricted by a ban on the manufacture and use of specific substances under the "Act on the Evaluation of Chemical Substances and Regulation of Their Manufacture, etc." and the "Agricultural Chemicals Control Act", among others.

The emission burden refers to the pollution that enters public bodies of water. The effluent standard in the Water Pollution Control Law regulates this emission pollution burden by regulating the concentration of effluent that enters public bodies of water. Furthermore, control of the total volume of water is implemented in landlocked bodies of water such as lakes, ponds, swamps, marshlands, and inland bays where regulation of pollutant concentration is inadequate, and the total volume of the inflow burden becomes a problem.

8.2 Standard Items for Preservation of the Aquatic Environment

8.2.1 Environmental Quality Standards

When an administrative agent takes measures for water preservation, as a first step, a standard is required to ascertain if pollution exists in the subject water. The purpose here is to decide if there is necessity for taking countermeasures. The Environmental Quality Standards was established for that reason. The Environmental Quality Standards is a standard, "the maintenance of which is desirable for the protection of human health and the conservation of the living environment" and is stipulated in Article 16 of the Basic Environment Law. This is the administrative goal to be achieved, and all environmental measures are implemented according to this.

There are Environmental Quality Standards concerning human health protection (health items) and environmental items concerning preservation of the living environment (living environment items).

Toxic substances harmful to human health include items such as heavy metals and agricultural chemicals, and nationwide uniform values are established as standards. That is to say, it is inappropriate to alter the standard value depending on the water or to exempt them from application in some other water, because human health must be respected above all. By the way, these standard values are roughly the same values as in the water quality standards for potable water based on the Water Supply Act.

Water parameters such as pH, COD or BOD, and SS are included in the living environment items. The Environmental Quality Standards applying to these are not set as nationwide uniform values. Instead, classes are set (classification) by the Minister of the Environment or by governors of prefectures, depending on the type of public body of water (e.g., rivers, lakes, ponds, marshland, or sea) and the use of such water.

Modification of standard values, and addition of standard items, has been done according to improved scientific judgment, and Environmental Quality Standards have been modified over time. Any changes in the Environmental Quality Standards health items reflect the water environment and focus on the society at the time of its establishment. Health-related items in 1970 included toxins that were considered to be the source of four major pollution incidents of the time (cadmium, cyanide, lead, hexavalent chromium, arsenic, mercury, and alkylated mercury). PCB was added in 1975, because it was the root cause of the Kanemi rice oil disease incident. Subterranean water pollution by cleaning facilities and agricultural chemicals from golf courses became problems in 1993, and 15 items were added to the list of concern, including VOCs (such as dichloromethane) and agricultural chemicals (such as thiuram). The Environmental Quality Standards for subterranean water were established in 1997 and adopted the same items and standards as the water Environmental Quality Standards for public bodies of water. Nitric acid pollution in subterranean water became a problem in 1999, and nitrate nitrogen and nitrite nitrogen were added to the lists, as well as fluoride and boron.

In 2009, 1,4-dioxane for public bodies of water and 1,4-dioxane, vinyl chloride monomer, and 1,2-dichloroethylene (sum of cis-isomer and trans-isomer replacing traditional cis-1,2dichloroethylene) for subterranean water were added. The last two items presuppose the existence of degradation products in subterranean water. Recently, different items tend to be adopted for public bodies of water and subterranean water, and the list of standard items is increasing in length.

8.2.2 Monitoring-Required Items

"Monitoring-required items" are for substances related to the protection of human health but are not required to be Environmental Quality Standards items immediately, judging from their levels in public bodies of water. It has been decided that efforts should be made to accumulate information.

The substance 1,4-dioxane was upgraded from a monitoring-required item and added to the list in the Environmental Quality Standards in 2009.

Research on monitoring-required items is conducted based on water measurement plans made by prefectures. The results are summarized by the national government, which then makes a public announcement. There are 26 inspection items as of 2014.

8.2.3 Investigation-Required Items

"Investigation-required items" are for substances for which the "water environmental risk" is relatively slight or unknown. However, accumulation of information concerning their "water environmental risk" is necessary from the viewpoint of detection in the environment or of combined effect. To date, 300 substance groups have been selected. The Ministry of the Environment conducts research projects before such substances are categorized as monitoring-required items, examining analysis methodology as well as concentrations and conditions.

8.2.4 Environmental Quality Standards for Preservation of Aquatic Life

Recently Environmental Quality Standards are being adopted that have not only humans as their subject but also aim at preservation of aquatic life forms (e.g., fish, crustaceans) which have a close relationship to human life. This type of Environmental Quality Standards concerning preservation of aquatic life (hereinafter referred to as Environmental Quality Standards concerning aquatic life preservation) is not treated as a health item but as a living environment item.

Environmental Quality Standards for aquatic life preservation are established using the following methods. First, preservation target values are calculated in consideration of the chronic toxicity of aquatic life forms and their prey and of subject substances. A lower concentration is chosen as the preservation target value of the subject substance. Meanwhile, the detection status in the actual environment and the preservation target value are compared, in order to select those for which excess is confirmed, which are then adopted as Environmental Quality Standards.

Environmental Quality Standards for aquatic life are classified according to their habitat (freshwater, seawater), preferred water temperature and preferred prey (e.g., cold water: char, salmon, trout; warm water: carp, crucian carp—this division applies to freshwater organisms only), and the living environment (general habitat and special area for spawning, young fry).

Total zinc was adopted as an Environmental Quality Standards item for aquatic life preservation in 2003, followed by the addition of nonylphenol in 2012 and LAS (linear alkylbenzene sulfonate) in 2013. In addition, chloroform, formaldehyde, and phenol were adopted as monitoring-required items for aquatic life preservation in 2003. The substances, 4-t-octylphenol, aniline, and 2,4-dichlorophenol were added to monitoring-required items in 2013. Currently, ammonia, copper, and nickel are under examination as candidates for addition to the list.

8.2.5 *Effluent Standard*

When the national government establishes Environmental Quality Standards, effluent control based on the Water Pollution Control Law is conducted as a guarantee. The standard for effluent control is the effluent standard. The effluent standard adopts values roughly ten times higher than the Environmental Quality Standards. This is because of the assumption that effluent is diluted about ten times with river water, at points with ample distance from the discharge opening, by the time it is discharged into public bodies of water.

Standard values of effluent standard health items are not set with a daily average but a maximum value. The assumed management (enforcement) response is direct punishment for effluent standard violations. The standard is established in both of the maximum value and daily average values for standard values of effluent standard living environment items. This is because daily average values can cause problems in water pollution concerning living environment items.

The effluent standard is set uniformly for all public bodies of water. However, further reduction of the pollution burden is necessary when the pollution burden is not low enough to achieve or maintain the Environmental Quality Standards. In such cases, prefectures can set more rigorous effluent standards (additional effluent standard) as prefectural ordinances that can be applied instead of the uniform effluent standard of Japan. In Fukuoka Prefecture, additional standards are set for the water of the Seto Inland Sea, the Omuta area, Hakata Bay, Onga River, Chikugo River, Yabe River, and the Chikuzen Sea. Table 8.1 shows the establishment of additional effluent standards in the water of the Omuta area in Fukuoka Prefecture.

Table 8.1 Additional effluent standard concerning water in the Omuta water area (Environmental Policy Section 2013)

1 The Omuta River (excluding the Omuta River bay area, the same shall apply hereinafter) and public bodies of water which flow into this river

Industry (facility)	Items and substances, and their allowance (unit mg/L)										Applied date
	BOD	SS	n-Hex (mineral oils)	Phenols	Cyanide compounds	Organophosphorus compounds	Cd, Cd compounds	Pb, Pb compounds	Cr^{6+} compounds	As, As compounds	
Human excreta treatment facility	45 (30)	120 (90)									
Other facility	15 (10)	100 (70)	1	1	Not to be detected		0.01	0.1	0.05	0.05	

2 Sea area of 500 m radius centering on the middle of the mouth of the Omuta River, also surrounded by the shore (including the Omuta River bay area), and public bodies of water which flow into this sea area (excluding the Omuta River and public bodies of water which flow into this)

Industry (facility)	Items and substances, and their allowance (unit mg/L)						Applied date
	BOD	COD	SS	n-Hex (mineral oils)	Phenols	Cd, Cd compounds	
Treatment facility of water discharged from specified workplace (excluding water which flows into public bodies of water) (excluding human excreta treatment facility and sewerage final treatment facility)		120 (100)	100 (70)	3	1	0.01	
Other facility	40 (30)	40 (30)	100 (70)	1	1	0.01	

3 Omuta water area excluding 1 and 2

Industry (facility)	Items and substances, and their allowance (unit mg/L)					Applied date
	BOD	COD	SS	n-Hex (mineral oils)	Phenols	
3-1 Specified workplace located in sewerage developed area						
All industries	30 (20)	30 (20)	100 (70)			
3-2 Existing specified workplace located outside sewerage developed area (workplace which had a facility equivalent to specified facility or had started constructing one as of April 1, 1973)						
Nonferrous metal manufacturing		40 (30)	100 (70)			
Mining and coal washing	70 (50)	70 (50)				
Slaughtering	120 (90)					
Human excreta treatment facility (limited to combined wastewater treatment)	45 (30)		120 (90)			
Animal-based feed manufacturing	120 (90)		100 (70)			
Sewage final treatment facility	30 (20)		100 (70)			
Other facility	120 (90)	120 (90)	150 (120)			April 1, 1990–
3-3 New specified workplace located outside sewerage developed area (workplace which had a specified facility (including those equivalent to one) or was going to be equivalent to one after April 2, 1973)						
Human excreta treatment facility	45 (30)		120 (90)			
Additional designated facility	120 (90)	120 (90)	150 (120)			April 1, 1990–
Other facility	30 (20)	30 (20)	100 (70)	1	1	

Values without parentheses are maximum and values in parentheses are average

As the table shows, additional effluent standards are set in detail, according to the water conditions where the effluent flows in, whether the facility discharging effluent (specified business operators) is located in an area with a sewerage system installed or not, and disregards business type.

8.3 Measures for Pollution Burden According to Pollution Source

Pollution sources are generally divided into point source and non-point source (also called area source). The point source can be effluent from a factory/workplace and/or household wastewater. Measures for workplace and household wastewater which are actually conducted in Fukuoka Prefecture are introduced here. The measures used for area source pollution are also described.

8.3.1 Measures for Factory/Workplace Pollution Burden

Facilities that emit effluent into public bodies of water are called specified facilities, and workplaces that own specified facilities are called specified business operators. Effluent regulation stipulated in the Water Pollution Control Law is shown in Fig. 8.2.

This law requires permit application for installing/modifying specified facilities and compliance with the effluent standard. When a permit is applied for installation/modification of a specified facility, the administrative agents carry out a rigorous check of the contents. Figure 8.3 shows the chronological change of the number of specified facilities in Fukuoka Prefecture. Most of the specified facilities were registered about 6–7 years after the Water Pollution Control Law was put into effect in 1970. After that, the number of specified facilities remained stable for some time, but since 2003 it has been on the decline. Nevertheless, there are still over 5,000 specified facilities in existence. However, the effluent standard that will

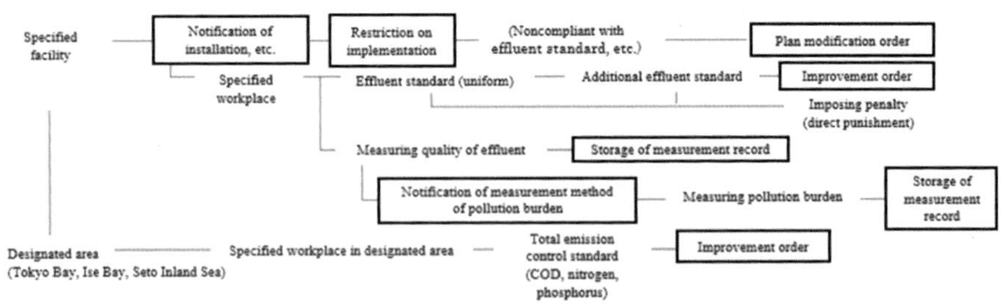

Fig. 8.2 Flow of effluent control for specified workplace (Japan Environmental Management Association for Industry 2013) partially altered

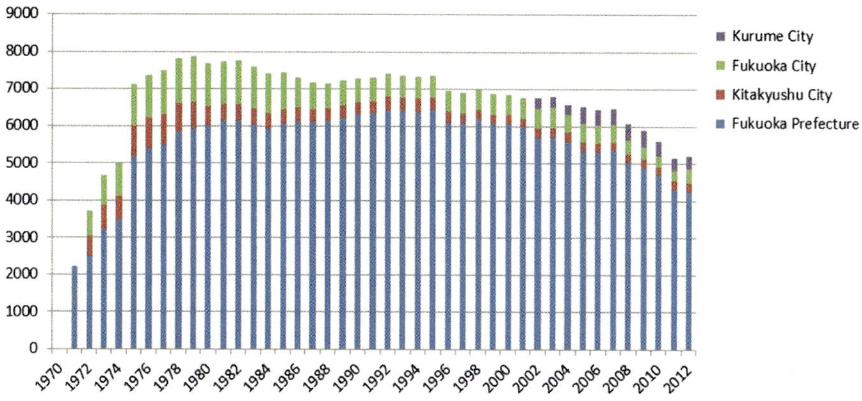

Fig. 8.3 Change in the number of specified facilities in Fukuoka Prefecture

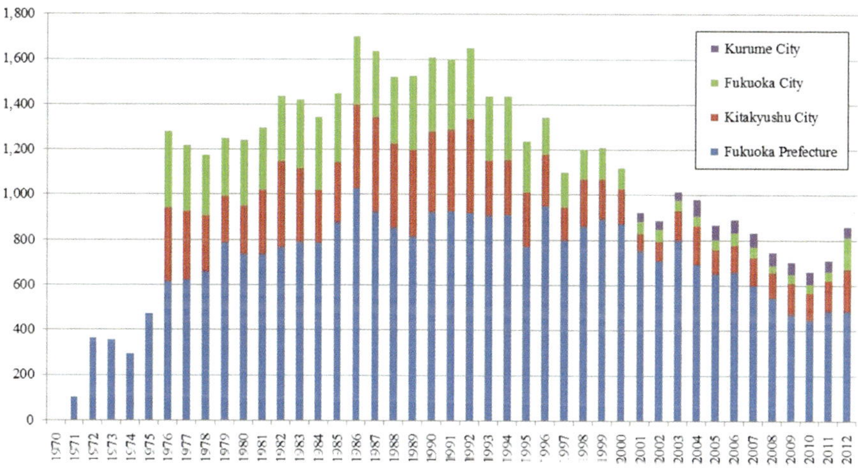

Fig. 8.4 Change in the number of on-site inspections at specified workplaces

be described later does not necessarily apply to all of these specified facilities. The effluent standard is only applied to facilities where the amount of effluent exceeds 50 m^3/day or where toxic substances are used. This did not apply to 79.8 % of specified workplaces as of the end of 2012. These are referred to as small workplaces, for which Fukuoka Prefecture gives improvement guidance for effluent water quality in the form of the "Fukuoka Prefecture Small Workplace Effluent Water Improvement Guidelines."

For those cases where the effluent standard is applied among specified workplaces, on-site inspection is conducted to confirm the conditions. The result of the inspection becomes the basis for taking necessary administrative measures. On-site inspection at a specified workplace in the prefecture is conducted by ordinance-designated cities and special cities in addition to the prefecture. Figure 8.4 shows the change in the number of on-site inspections at specified workplaces in Fukuoka

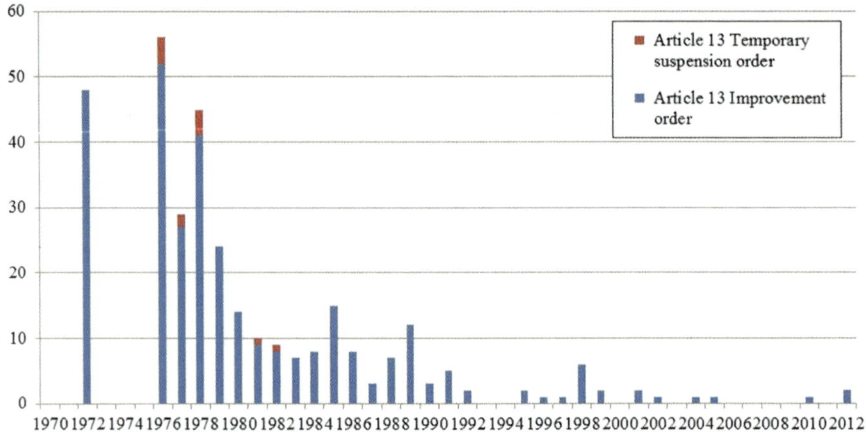

Fig. 8.5 Change in the number of improvement order/drainage suspension orders issued to specified workplaces

Prefecture. Over 1,600 cases of inspection were carried out at its peak soon after the Water Pollution Control Law was put into effect. However, the number has been falling in recent years; a total of 700–800 cases of on-site inspection are now carried out in workplaces annually.

Administrative guidance is given to workplaces which do not satisfy the effluent standard. Administrative agents give instruction, advice, and suggestions to business operators so that they will be able to comply with the effluent standard, and it is not an administrative disposition. If measures are not taken at workplaces and no improvement is shown, an improvement order is issued, based on Article 13 of the Water Pollution Control Law. If the improvement order is not complied with, a drainage suspension order is issued. A drainage suspension order necessarily leads to shutdown of business, a matter of life and death for business operators. Figure 8.5 shows the change in improvement orders/drainage suspension orders issued to specified workplaces in Fukuoka Prefecture. As many as 50 improvement orders, or more, and four drainage suspension orders were issued annually for 10 years after the enactment of the Water Pollution Control Law. In recent years, the number of these orders issued has fallen markedly. Nowadays, there are as few as two improvement orders, if any, every year. As for drainage suspension orders, no such order has been issued in the past 30 years, which means agreeable conditions are continuing.

Direct punishment is stipulated for violating the effluent standard in the Water Pollution Control Law. Figure 8.6 shows the change in the number of penalties applied to specified workplaces in Fukuoka Prefecture. In the 1970s, when pollution was a problem, over 25 penalties were applied. Pernicious violations were exposed by the prefectural police. Compared with those days, we can see there has been a considerable decrease in the issuance of improvement orders, business suspension orders, and penalties. This implies that measures for water quality preservation at workplaces are effective because the number of administrative orders and penalties are decreasing, despite the fact that there are a certain number of specified facilities in existence.

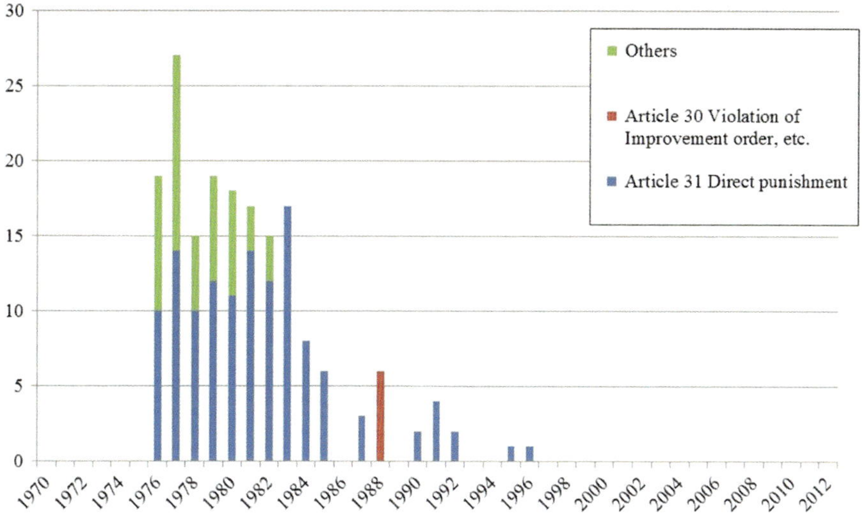

Fig. 8.6 Change in the number of penalties imposed on specified workplaces in Fukuoka Prefecture

8.3.2 Measures for Domestic Wastewater Burden

Compared with the time when there were serious traditional pollution problems, the proportion of effluent coming from workplaces is becoming smaller. As previously mentioned, this is because measures at workplaces have been advanced or economizing water consumption or water recycling at workplaces has been improved. On the other hand, the proportion of household wastewater in the pollution burden is becoming greater, and reduction of this burden is an important agenda for water preservation.

In Fukuoka Prefecture, areas where there is a low Environmental Quality Standards achievement rate are designated as intensive household wastewater measures areas. Municipalities that are designated as intensive household wastewater measures areas make a "household wastewater measures promotion plan," develop sewerage systems and domestic sewage treatment tanks, and implement enlightenment activities. The proportion of the population served by a sewerage system in Fukuoka Prefecture as of 2011 was 77.4 % higher than the national average of 75.8 %. However, the proportion goes down to 57.2 % when the two ordinance-designated cities of Fukuoka and Kitakyushu are excluded. It cannot be said that there is adequate sewerage development in provincial cities (Sewerage Section 2013). In the future, the prefecture aims to reduce the household-type pollution burden by implementing these measures and improving the Environmental Quality Standards achievement rate as well.

8.3.3 Measures for Pollution Burden of Non-point Source

As was previously mentioned, the pollutants originating from workplaces and households has clear inflow points, which are called point sources. As measures for point sources, in other words, measures for workplaces and households, are developed, the rate of non-point-source pollution increases. Non-point-source pollution takes place over a wide area, and it is more difficult to control compared with point-source pollution. This has become a problem.

For example, the outflow of pollutants in rain from forests, farmland, and towns is non-point-source pollution. Measures may be taken, such as woodland improvement in forests, covering bare soil and suspending open freight storage of fertilizer on farmland, and permeable pavement and cultivation in towns. However, the problem is that we cannot see the effect quickly and clearly enough because the subject area of the measures is large.

The policy of basic measures for non-point-source pollution is that suppressing burden generation should be the basis rather than emitted burden or runoff burden. In addition, approximate examination of size and combination of measures should be conducted before detailed examination. Moreover, areas where the effect is considerable from the result of approximate examination should be specified, and appropriate measures taken (Ministry of Land, Infrastructure, Transport and Tourism et al. 2006).

8.4 Reinforcing Water Purification Capacity

8.4.1 Measures for Reinforcing Water Purification Capacity

The concentration of pollutants in rivers tends to diminish as the water flows downstream due to decomposition, sedimentation, or adsorption. Generally speaking, the degree of purification in rivers is indicated as a percentage of retention of purification, or a self-purification coefficient. For example, the apparent self-purification coefficient is calculated even when the subject substance is a conservative material (i.e., it does not decompose like organic substances because it decreases in water due to the abovementioned actions). Therefore, we need to have a clear view of the behavior of substances in water in order to take effective measures, making use of the water purification capacity. In other words, we need to understand the elementary processes that dominate the material cycle in water. Table 8.2 shows methods of river purification measures adopted in the field (Kusuda 1994).

8.4.2 Cases of Reinforced Water Purification Capacity

Fukuoka Prefecture has developed a special type of concrete block containing bamboo charcoal (Tsuchida et al. 2004), which is used for river bank protection

Table 8.2 Various purification measures in rivers (Kusuda 1994)

Measures by changing shape, amount of flow	Measures by purification facility
Retarding basin, reservoir	Coagulating sedimentation method
Inducing water for purification (water for maintenance)	Installing nonwoven fabric precipitation membrane
Altering flow	Rapid filtration method
Riverbed development	Filtration by long-fiber filter method
Dredging earth and sand	Micro-strainer
Removing bottom sediment	Permeable waterway
Inducing bottom wetland	Soil treatment (adsorption on soil)
Reinforcing subterranean accumulation, permeation	Direct aeration
Subterranean dam, subterranean water cultivation	Contact oxidation method
Hydrophilic embankment (planting)	Gravel contact oxidation method
Installing movable weir	Oxidation pond (OD method)
	Preparing reed bed

in order to increase the purification capacity of rivers. It does this by using the contact oxidation function. This technique has the additional merit in that it provides an effective use for long-neglected bamboo thickets throughout the nation that are becoming problematic. Porous bamboo charcoal is mixed into the concrete block so that microorganisms attach themselves to its surface, which reinforces water purification. When compared with ordinary concrete block for BOD, bamboo charcoal registered a faster reduction. Also approximately three times more microorganisms attach themselves than on ordinary concrete block, showing the increased removal effect of BOD. Currently construction sites are in place on the Umi and Hiyama Rivers.

8.5 Current Situation and Future Problems in Fukuoka Prefecture

8.5.1 Constant Monitoring of Water at Environmental Quality Standards Survey Points

Environmental Quality Standards survey points are established in each body of water to grasp the situation of pollution in regard to the Environmental Quality Standards there. Furthermore, in addition to these Environmental Quality Standards survey points, constant monitoring of water is conducted at Environmental Quality Standards supplementary survey points. Table 8.3 shows the implementation situation of constant water monitoring. In Fukuoka Prefecture in FY2013, a total of 4,350 constant water monitoring activities were carried out at 458 points, in

Table 8.3 Implementation condition of constant water monitoring in Fukuoka Prefecture (Environmental Policy Section FY 2013)

Division		Prefecture		National Government (Ministry of Land, Infrastructure, Transport and Tourism, Japan Water Agency)		Ordinance-designated city, core city		Other municipality		Total	
Public bodies of water survey	Survey division	Number of points	Total number of inspections	Number of points	Total number of inspections	Number of points	Total number of inspections	Number of points	Total number of inspections	Number of points	Total number of inspections
	Rivers	80	880	28	304	104	834	181	960	393	2,978
	Sea	19	384	–	–	29	712	–	–	48	1,096
	Lakes and reservoir	9	132	7	108	1	36	–	–	17	276
	Total	108	1,396	35	412	134	1,582	181	960	458	4,350

cooperation between the prefecture, national government, ordinance-designated cities, core cities, and other municipalities (Environmental Policy Section 2013).

8.5.2 Result of Constant Monitoring of Water

The following is the result of constant monitoring of water in Fukuoka Prefecture in FY2012 (Environmental Policy Section 2013).

As for Environmental Quality Standards health items, the Environmental Quality Standards was cleared in all the public bodies of water apart from those points where fluoride and boron exceeded the standard due to the effect of sea water.

As for Environmental Quality Standards living environment items, the achievement rate of representative indices of organic pollution BOD (in rivers) and COD (lakes and reservoirs, sea) was 82.8 % in total public bodies of water. The breakdown was 81.8 % in rivers, 87.9 % in the sea, and 80.0 % in lakes and marsh/ponds.

8.5.3 Improved Water Quality in Fukuoka Prefecture

Figure 8.7 shows the chronological change of water quality in key bodies of water in Fukuoka Prefecture in the 40 years from 1971 to 2011. BOD in rivers and COD in the sea are used as indices of organic pollution.

Since the period of rapid economic development in the 1950s–1960s, industrialization and population concentration increased, especially in the metropolitan areas of Kitakyushu, Fukuoka, and Omuta. Water pollution increased right after the enactment of the Water Pollution Control Law in FY1971, but the water quality in key bodies of water in Fukuoka Prefecture has been greatly improved thanks to various measures at the pollution sources, as was previously mentioned.

For example, the Omuta River, which flows in the center of Omuta City, had problems regarding health and impact on the fishery due to offensive odors, dye, and toxic substances. This was because effluent from chemical works, metal refining facilities, and mines on the basin had been allowed to flow without treatment. When we look at the approx. 100,000 tons of water flowing in the Omuta River at that time, ~ 80,000 tons was industrial effluent, ~ 20,000 tons was urban wastewater, and the volume of natural water was as little as 2,000–3,000 tons. The river resembled a factory drain (Pollution Control Section 1971). As a result, the BOD at that time registered an extremely high figure of 256 mg/L. Thanks to effluent regulation, including an additional effluent standard in the area, it is now approaching the water level of an ordinary river.

Moreover, the increase of BOD in the midstream area of the Mikasa River in the 1970s–1980s was caused by an increase in household wastewater accompanying the rapid expansion of residential areas in Metropolitan Fukuoka (Environmental and Fukuoka Prefecture 1972), but the water quality after that has been improved thanks to the development of a sewerage system.

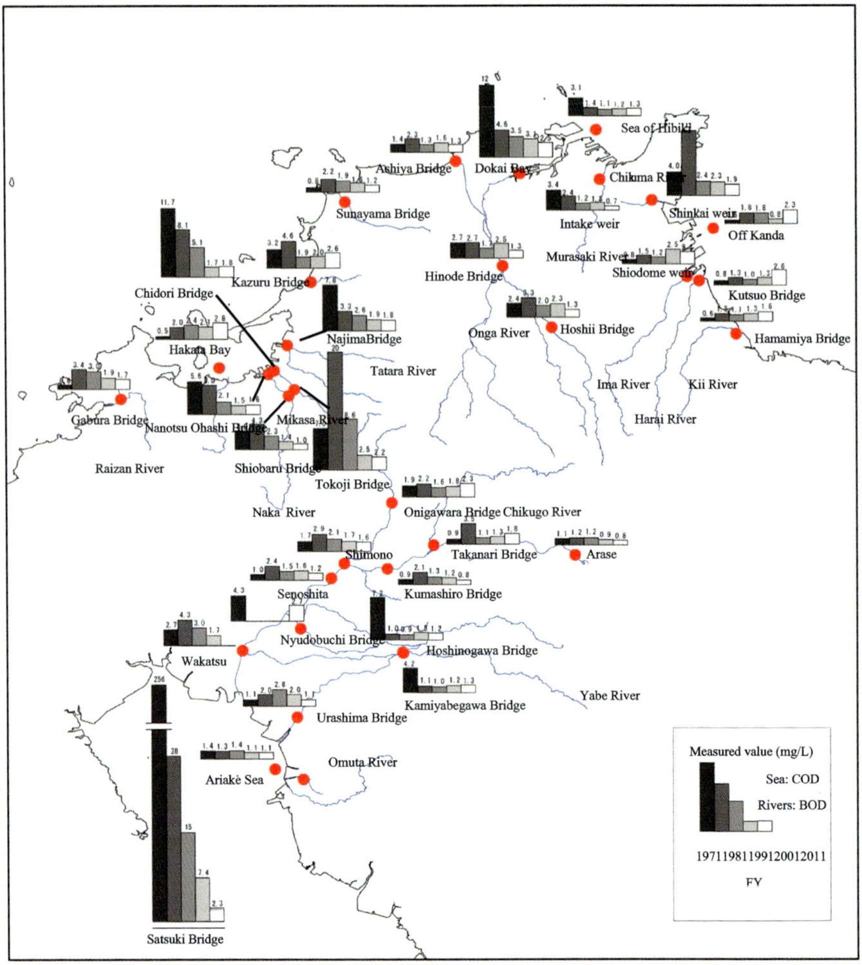

Fig. 8.7 Change of BOD/COD concentration in key bodies of water of Fukuoka Prefecture over 40 years (Note: Data at Satsuki Bridge in 1972 is used for data at Sunayama Bridge in 1971. Data at Kaneshima Bridge have been used since 1991 for Tokoji Bridge. Data at Benten Bridge in 1972 is used for data at Gabura Bridge in 1971. Data at Shintani Bridge in 2011 is used for data at Nyudobuchi Bridge in 2011. Data for Buzen Sea S-2 has been used since 1981 for data off Kanda D-9. C-1 data has been used for central Hakata Bay St7 since 1981. Data in 1972 is used for data at Hoshii Bridge, Shiodome weir, Kutsuo Bridge, Hamamiya Bridge, Kazuru Bridge, and Shimono in 1971)

8.5.4 Future Problems Concerning Preservation of the Aquatic Environment

As was previously described, the number of Environmental Quality Standards items and monitoring-required items are increasing every year, and it is expected to increase in the future as well. On the other hand, local authorities that have been in charge of constant monitoring of water in the public bodies of water for a long time are now faced with business reduction due to tightness of local budgets. This has led to reductions in the number of staff in charge and in the budget for environmental preservation, replacement of outdated analysis equipment, and loss of analytical skills and knowledge of technology with the departure of aging staff (Iwamoto 2009). All of this increases the work burden of constant water monitoring. In order to carry out accurate and efficient water monitoring work with limited budgets and manpower under such circumstances, the Ministry of the Environment drew up the "Guideline for optimizing and emphasizing water measurement concerning establishment of measurement plan for public bodies of water" (Environment Management Bureau and Ministry of the Environment 2009) and suggests reducing the number of water monitoring points and decreasing the frequency of monitoring in areas which do not have a history of water quality problems. Development of environmental analysis technology is also becoming necessary, for such as developing a qualitative screening method effective for energy saving, and a multipurpose simultaneous analysis method that includes pretreatment.

Meanwhile, outsourcing of sampling/analysis work previously done by local administrative agencies to private institutions is on the way (Iwamoto 2009). Under such circumstances, the importance of measures to secure the credibility of measurement by analysis agencies and to aim for improved accuracy (e.g., by a survey of uniform accuracy management for environmental measurement analysis, which has been conducted by the Ministry of the Environment since 1975) is increasing.

References

Environment Management Bureau, Ministry of the Environment (2009) Guideline for optimizing and emphasizing water measurement concerning establishment of measurement Plan for Public Bodies of Water

Environmental Bureau, Fukuoka Prefecture (1972) Annual report on environment FY1972

Environmental Policy Section, Department of Environmental Affairs, Fukuoka Prefecture (2013) Annual report on environment FY2013 edition

Iwamoto S (2009) An observation on how local environmental laboratory should be. J Environ Lab Assoc 34:34–39

Japan Environmental Management Association for Industry (2013) New/technology and laws for pollution prevention 2013, on water quality, Tokyo

Japan Sewage Works Association (2008) Survey of comprehensive sewerage improvement plan for individual river basin guiding principles and explanation, Tokyo

Kusuda T (author and editor) (1994) Reinforcement and control of natural purification mechanism, Gihodo Shuppan, Tokyo

Ministry of Land, Infrastructure, Transport and Tourism et al (2006) Basic idea for measures on the basin for water quality in lakes and ponds—measures for burden from non-point source

Pollution Control Section, Department of Public Health and Medical Affairs, Fukuoka Prefecture (1971) Annual report on environment FY1970

Sewerage Section, Department of Structures and Urban Planning, Fukuoka Prefecture (2013) Sewerage system in Fukuoka FY2012

Tsuchida D, Ishibashi Y, Tokunaga T, Seri K, Kuratomi S (2004) Characteristic of water purification by concrete block mixed with bamboo charcoal. J Environ Lab Assoc 29:102–106

Chapter 9
Recyclable Resources and Proper Waste Disposal

Takayuki Shimaoka

Abstract Economic and social activity promotes the generation of waste; in this respect, the text presented here discusses problems associated with waste in Japan and the efforts that have been made to solve them. Additionally, the text introduces technical transitions that have occurred in relation to waste disposal and final disposal. The following is an overview of the flow of materials in Japan, and it describes the amount of natural resources used, materials reused, and waste generated, thus providing ideas for use in policies related to building a recycling-oriented society. The stock amount of materials has now surpassed 30 % of the total amount of materials input, and materials supplied both domestically in Japan and internationally are accumulating. Although it is acknowledged that this stocked material could potentially be transformed into massive amounts of disaster waste in an instant if a large natural disaster occurred, it is considered that such material also has the potential of being used as a precious recyclable resource. This text discusses the move toward waste-based recyclable resources, which are aimed at creating a sustainable society. It delves into sustainable environmental technology used to turn large amounts of waste of varying quality into recyclable resources and also introduces an example of sustainable environmental technology.

Keywords Waste disposal • Intermediate processing • Final disposal • Recycling-oriented society • Resource recycling • Sustainable recycling technology

9.1 History Associated with Waste Problems and Measures Taken to Solve Related Issues

The following is an outline of the waste problems experienced in Japan and the measures taken to correct such problems, which are inevitably related to economic and social activity and the associated actions of people living their lives. It is considered that certain types of waste that have been disposed in landfill over the years could possibly be reused as resources, and the development of technology to

T. Shimaoka (✉)
Department of Urban and Environmental Engineering, Faculty of Engineering,
Kyushu University, Fukuoka, Japan
e-mail: shimaoka@doc.kyushu-u.ac.jp

T. Shimaoka et al. (eds.), *Basic Studies in Environmental Knowledge, Technology, Evaluation, and Strategy*, DOI 10.1007/978-4-431-55819-4_9

enable such a process is currently being demanded. In addition to discussing this, the text also summarizes the history of the move toward recycling-focused waste disposal and the shift toward using waste as a resource.

Society requires waste disposal as a public health measure, and this has been an important social issue throughout history. For example, waste was a major social issue in the Edo period, and a proclamation banning open-air burning of household garbage was issued in 1655. During this era (1661–1672), waste was transported to Eitaijima for final disposal by specialty contractors, where it was disposed of. Landfill operations at Eitaijima ceased in 1730, and waste was gradually transferred to Etchujima, Fukagawa ((Foundation) Suginami Shoyo Memorial Foundation 1983).

Moving into the Meiji period, a law regarding disposal of household garbage was enacted in 1888. As foreign culture began to be introduced in the Meiji period, infectious diseases such as cholera also became prevalent. Cholera was rampant from 1895 to 1896, with the number of casualties reaching tens of thousands. As part of the efforts to combat this disease, the Meiji government declared it mandatory to bring the disposal of garbage into the public sector in 1900, enacting the Public Cleansing Act. This was Japan's first law related to waste management.

In 1897 during the Meiji period, a furnace capable of incinerating 11.5 t of garbage daily was constructed in Tsuruga City, with an attached chimney measuring 1.8 m high; this represented the first use of an incinerator in Japan. Two years after the Great Kanto earthquake, a waste incineration plant was completed in Osaki, Tokyo, in 1925 and was constructed with a furnace capable of burning 22.5 t of garbage in 10 h (Mizoiri 1988). Moving into the Showa period, the number of waste incineration plants in Tokyo increased to eight facilities between 1927 and 1936; 45 % of garbage was disposed of by incineration and 45 % as land fill, while 10 % was used in fertilizer and feed.

Following World War II, human waste was no longer used as a fertilizer in farming villages due to the rapid spread of chemical fertilizers and the huge transformation of farming villages triggered by agricultural land reforms. As it was no longer of use, the treatment of human waste became an issue. During this time period, waste disposal became a public health issue in order to maintain a sanitary and pleasant living environment. The Public Cleansing Act was enacted in 1954 and the Public Cleansing Act was abolished (Japan Waste Management Association 1996).

The current waste problem began in 1960, when Hayato Ikeda, the Prime Minister at the time, advocated for income doubling and lauded a society in which consumption was a virtue. As income rapidly increased, the amount of waste simultaneously ballooned, and by 1970 over 1 kg of waste was being produced per person daily. With the demand for labor in the 1960s, the population also began to be concentrated in cities, and with waste policies unable to keep up with the abrupt expansion of cities, places like marshlands and unused reservoirs were chosen as makeshift landfills. Subsequent historical matters of note related to waste are as described below:

1957: Sea surface disposal of waste from 23 wards of Tokyo began at Site 14 (commonly known as Yumenoshima).

1965: Flies appeared in huge numbers at Yumenoshima, Tokyo. Numbers were so large that the roofs of houses in neighboring Koto-ku looked completely black.

1940: In the December session of Diet, 14 bills were passed in relation to pollution, and this session is widely referred to as "the pollution Diet." The Waste Management Act was passed in the same year, with waste normally referred to as garbage designated as general waste and waste generated by industry categorized as industrial waste.

1971: In response to resident activism relating to the waste disposal sites at the tip of the Koto-ku area, Tokyo Governor Minobe declared a war on the waste of Tokyo at the Tokyo Metropolitan Assembly and began working on the waste problem. This was the catalyst for an increase in social awareness of landfill disposal. The Environment Agency was also founded this year.

1972: Construction of Japan's first fluid bed incinerator was completed in Matsudo. Two years later in 1974, Japan's largest incinerator handling 1,800 t per day was created in Koto-ku, Tokyo.

1977: Structure Guidelines of Waste Disposal Facilities were established for the first time in Japan, while at the same time, establishment of seepage control was clarified by a joint order.

1978: The Osaka Bay Phoenix Project, a final disposal site plan utilizing the sea surface in a metropolitan area (a regional disposal development plan for the Osaka Bay area), was announced. Four years later in 1982, the Osaka Bay Regional Offshore Environmental Improvement Center was established.

1979: In Kamaishi, Nippon Steel constructed two 50 t waste-melting furnaces that used the principles of melting furnaces. In the 1980s, research and development on gasification melting technology began as part of research on recycling technology.

1983: The magazine *Kurashi no Techo* publicized the issue of mercury leakage from dry-cell batteries in landfill, sending shock waves through society. Furthermore, other research disclosed pollution from heavy metals and chlorine-based chemicals (such as the publication of 1983 research by Professor Tachikawa of Ehime University on the dioxin content of incineration residue), which was another shock disclosure.

1986: The leachate leakage problem, created by damaged liner system at a landfill site in Tobuki, Hachioji, invigorated the public's anti-landfill movements.

1989: The Japan Society of Waste Management was established in 1990 following the establishment of the Japan Waste Management Research Foundation, a move long-awaited by people involved in waste management.

1991: With the passing of the Law for the Promotion of Utilization of Recycled Resources, the use of waste as a resource was legally established for the nation. This was the first major legal amendment since the enactment of the Waste Management Act in 1970, and it established a system for use with specially controlled waste. At the same time, the Act on the Promotion of Development of Specified Facilities for the Disposal of Industrial Waste was promulgated.

1993: The problem of illegal dumping in Teshima, Kagawa Prefecture, finally became a national issue with the national Environmental Dispute Coordination Commission moving to intervene. In this year, the Act on Special Measures concerning Removal of Environmental Problems Caused by Specified Industrial Wastes (Industrial Waste Special Measures Act) was promulgated.

1995: The Act on the Promotion of Sorted Garbage Collection and Recycling of Containers and Packaging (Containers/Packaging Recycling Act) was promulgated; this was the first time waste resources were used effectively in Japan. It is of note that the Great Hanshin earthquake generated approximately 20 million tons of disaster waste.

1997: Countermeasure notices and new guidelines for reducing dioxins from waste disposal were formulated, and a movement began to lower the concentration of dioxins in the air to the worldwide average. One suggested countermeasure was to begin decreasing the number of small batch furnaces (which represented a number of garbage incineration facilities) handling general waste. At the same time, the establishment of facilities performing operations such as the melt solidification of incineration residue was promoted.

1998: The Act on Recycling of Specified Kinds of Home Appliances (Home Appliance Recycling Act) was promulgated, and the collection of four kinds of large household appliances (televisions, washing machines, refrigerators, and air conditioning units) began.

2000: The Basic Law for Establishing a Recycling-oriented Society, a basic law for effectively using and recycling the resources of Japan, was promulgated. In addition, in this year, the Act on Recycling Construction-Related Materials (Construction Material Recycling Act) and the Act on Promotion of Recycling and Related Activities for Treatment of Cyclical Food Resources (Food Recycling Law) were promulgated. The Act on Promoting Green Purchasing, which supported these laws, came into force in 2001. In 2002, the Act on Recycling, etc. of End-of-Life Vehicles (Vehicle Recycling Act) was promulgated, and efforts geared toward forming a resource-recycling-oriented society advanced steadily. At the same time, final disposal sites, which are central to the proper waste disposal technology that bolsters the 3Rs, began to fulfill a vital role as city facilities that promoted sustainable conservation of the environment.

2001: The Act on Special Measures concerning Promotion of Proper Treatment of PCB Wastes was enacted, and it was made mandatory to complete the reliable and proper disposal of PCB waste by 2016.

2003: The goal of enacting the Act on Special Measures concerning Removal of Environmental Problems Caused by Specified Industrial Wastes was to deliberately and steadily promote the elimination of problems arising from industrial waste that was disposed of improperly or through illegal dumping. Following this, the period of validity was extended to 2023.

2009: After being proposed by the Japanese government, the Regional 3R Forum in Asia was founded on the premise that the population was increasing in Asia, urbanization was progressing, the amount of waste was skyrocketing, and the quality of waste was diversifying.

2011: The Great Tohoku earthquake generated approximately 20 million tons of disaster waste included tsunami deposit. The creation of radioactive contamination waste in relation to the Fukushima Daiichi nuclear disaster made it difficult to dispose of disaster waste and decontamination waste.

2013: To promote the recycling of items such as small postconsumer electronic devices including digital cameras and game consoles, the Act on Promotion of Recycling of Small Postconsumer Electronic Devices (Small Home Appliance Recycling Law) was enacted.

9.2 Changes in Waste Management Technology

9.2.1 Intermediate Treatment

(a) Types and characteristics of incineration technology

Socially recognized methods of waste disposal within Japan's limited territory are incineration and landfill. The goal of incineration is to achieve the following at a high resource efficiency and low cost: volume reduction, amount reduction, stabilization, detoxification, and recycling of waste; landfill life extension; and a reduction in transportation costs. Japan Machine Furnace No. 1 (15 t/8 h, smokestack 33 m) was constructed in 1960 in (what was at the time) Tamana, Okayama Prefecture. Following this construction, the generation of waste increased rapidly, and the demand for full continuous incinerators rose in the midst of the postwar economic boom emblematic of the Iwato economy (Mizoiri 1988). Osaka City was the quickest in the nation to plan for the construction of incinerators that could be operated continuously for 24 h, and the development of the Sumiyoshi Plant in 1963, which was a full continuous incinerator plant, was completed.

(b) Changes in exhaust gas disposal (Shigaki 2000)

The three full 150 t/24 h incinerators completed in 1963 at Osaka City's Sumiyoshi Plant were Japan's only fully continuous incinerators. During this time, environmental damage due to air pollution was becoming a social issue, and the Osaka incinerators used multicyclone and electric precipitation equipment for fly ash removal. In 1968, the Air Pollution Control Act came into effect in place of the Smoke and Soot Regulation Act. Instead of the emission standards previously in place to control sulfur oxide, maximum ground concentration (K value) regulations were put in place, and smokestacks fitted with nozzles became widespread. Osaka antipollution regulations were enacted in 1970, mandating that wet exhaust gas cleaning equipment should be installed to remove hydrogen chloride. When wet cleaning equipment subsequently became widespread, hydrogen chloride exhaust regulations were applied, and these were then applied to waste incinerators in 1977. In addition to wet cleaning, the hydrogen chloride exhaust regulations led to the development of dry exhaust gas treatment technology that was easy to install and could economically blow in slaked lime. Nitrogen oxide exhaust regulations were enacted in 1973, and the non-catalytic denitration method and the low-oxygen combustion control method were developed.

(c) Regulations such as dioxin countermeasures

In 1983, Professor Tachikawa of Ehime University published data on the dioxin content of incineration residues from an incineration plant, and guidelines for preventing the production of dioxins were formulated in 1990. In 1996, Japan set baselines for dioxin concentrations corresponding to the size of each incinerator (separated into new and existing facilities) and issued a notice stating that melt solidification was to become the basic method used for the treatment of incineration residues. Melting facilities therefore began to be constructed next to incineration facilities nationwide, and the use of melting to treat of residue left after incineration became a common practice. In 1997, the country announced countermeasures (new guidelines) to reduce dioxins from waste treatment, attempting to promote the creation and operation of melt solidification facilities. In the same year, part of the Air Pollution Control Act was amended, designating dioxins as hazardous air pollutants and subscribing emission standard values for them. Furthermore, the Act on Special Measures concerning Countermeasures against Dioxins was enacted in 1999, and the amount of dioxins subsequently emitted dropped year upon year.

Waste incineration facilities were the biggest source of dioxins, and emissions of dioxins from municipal waste incineration facilities were estimated at 5,000 g TEQ per year in 1997. However, this number fell to 71 g TEQ per year in 2003 due to the strengthening of regulations, thereby achieving the reduction goal established based on the Act on Special Measures concerning Countermeasures against Dioxins. The creation and operation of melt solidification facilities in place of incineration disposal were thus promoted as a countermeasure against dioxins.

(d) Melting technology for waste

The current facilities used for melting are either ash-melting facilities that melt incineration residue or facilities that directly melt waste. According to a fact-finding survey, the first direct-melting facility was established in Kamaishi, Iwate Prefecture, in 1979. Following this, a fuel-based ash-melting facility was built in 1985, and an electrical ash-melting facility in 1991, but there were then no further marked increases in the development of melting facilities, and by 1993 there were still fewer than ten facilities in existence in Japan. However, based on the Act on Special Measures concerning Countermeasures against Dioxins, new standards were applied to emissions of dioxins, taking effect in December 2001. Accompanying this, most local governments decommissioned existing waste incineration facilities by the end of 2002 or built new gasification/direct-melting furnaces or incineration facilities alongside ash-melting facilities.

(e) Specially controlled waste: PCB waste disposal

Triggered by the Kanemi rice oil disease incident in 1968, the production and use of PCBs (polychlorinated biphenyls) were terminated in 1972, and it became mandatory to store them. Up until this time, however, Japan had produced approximately 60,000 t of PCBs, and although the storing of PCBs produced over the past 30 years was therefore unavoidable, it became problematic to establish relevant disposal facilities. Furthermore, the Stockholm Convention on Persistent Organic Pollutants (POPs Convention) was adopted internationally in 2001, aiming to eliminate POPs such as PCBs by 2028. With these circumstances as a backdrop, the Act on Special Measures concerning Promotion of Proper Treatment of PCB Wastes (PCB Special Measures Act) was enacted in 2001, mandating that PCBs should be disposed of within 15 years. Based on the Waste Management Act, four types of chemical decomposition methods were sanctioned: dechlorination decomposition, hydrothermal oxidation decomposition, reduction thermochemical decomposition, and photo decomposition.

9.2.2 *Final Disposal*

(a) Final disposal site classification and landfill structure

In around 1965, final disposal sites were exclusively known as landfill sites, and this referred to an area of land that was filled in with waste. At this time, land other than that in large cities, such as mountain areas, marshlands, and reservoirs, was primarily used as landfill by dumping. However, these areas only began to be developed as final disposal sites after 1975.

The Waste Management Act was enacted in 1970, with waste classified as either industrial waste or municipal waste. Based on the type of waste, final disposal sites for industrial waste were divided into quarantined, controlled, and stabilized disposal sites. All final disposal of general waste falls into the controlled category (as classified for industrial waste). Controlled final disposal sites are used to dispose of industrial waste as land fill (if it is below the specified elution standard for heavy metals and hazardous materials) and waste with a dioxin content under 3 ng-TEQ/g, such as bottom ash and fly ash. Quarantined final disposal sites are used to dispose of industrial waste as landfill that does not meet these criteria. Stabilized final disposal sites are designated for industrial waste that does not interfere with environmental conservation, even when buried in its original form, such as glass, ceramic waste, and scrap rubber.

Landfill sites are classified as being areas of either land reclamation or water surface reclamation depending on where they are located, and the latter is further divided into sea surface landfill sites and freshwater surface landfill sites. Although sea surface landfill sites have been operating since the Edo period, at the time they were shore-based sites, as opposed to the current sites that are closed to the sea surface. Historically, land-based landfills were often situated in mountainous areas and were uncommon on flat land.

The United States and the countries of Europe have employed an anaerobic landfill structure, and the use of methane gas generated by landfills is widespread. However, due to climate conditions and societal demand, Japan uses an aerobic landfill structure, which is rarely used internationally. The semi-aerobic landfill structure came into use around 1975. It works by increasing the movement of aerobic microorganisms in relation to feeding air inside the landfill; the internal temperature reaches 60–70 °C, which is higher than the outside air temperature. As a result, the gas density inside the landfill decreases, and this is expelled into the air at the same time that air is also injected into the landfill through perforated pipes at the bottom of the landfill. Using this mechanism, the semi-aerobic structure actively brings air into the landfill and uses aerobic microorganisms to hasten improvement of the leachate quality and the decomposition of landfilled waste.

(b) Development of seepage control technology

1. Surface seepage control

   At final disposal sites for waste, precipitation soaks into the waste layers and becomes leachate, and this eventually passes through the leachate collection drain system at the bottom of the landfill and collects at a leachate treatment facility. Types of seepage control established with the goal of preventing groundwater contamination from this leachate can be broadly divided into either vertical impervious walls or surface liners. In a June 1998 standards ministerial ordinance revision regarding seepage control at final disposal sites, plans were made for the "Duplication of seepage control through a combination of seepage control liner sheet and impermeable soil, and a strengthening of the seepage control function through the creation of a protective layer".

2. Leakage detection technology for surface seepage control

   The problem of groundwater contamination due to leakage of leachate from landfill sites is the primary cause of opposition toward the construction of final disposal sites, and in this respect, there is demand from the management of final disposal sites for development of special seepage control leakage detection technology. Broadly speaking, there are two types of methods currently being developed for detecting and locating damage to seepage control sheeting: electrical detection methods and physical detection methods. Electrical detection methods include the potential method, the leakage current method, pulse method, current phase method, and impedance method, while physical detection methods include the vacuum suction method and the colloid solution pressurization method. These methods are making it possible to detect even extremely small amounts of damage.

(c) Closed final disposal sites that are easily accepted by the community

Due to past environmental issues with final disposal sites, public suspicion of these sites has risen, thereby damaging the progress of final disposal site construction. To assist in alleviating such suspicions, the idea of closed final disposal sites was conceived. The characteristics of these disposal sites are as follows:

From an outside perspective, the final disposal site gives an environmentally friendly impression.

The disposal site is closed off, and therefore waste cannot escape in the breeze and odors cannot be spread.

The amount of leachate produced is not affected by natural phenomena such as precipitation (rainfall, snowfall), and the amount can therefore be controlled.

There are further additional advantages, such as the ability to continue landfill operations even in snowy regions during the winter, making it easy for the community to accept such disposal sites (Closed System Disposal Site Development Association 2004). However, there continues to be a concern that technology required to stabilize the landfilled waste has not been adequately developed, and it is therefore considered that such operations should cease.

9.3 Transition to Waste-Based Recyclable Resources Aimed at Creating a Sustainable Society

9.3.1 The Modern Resource Problem

Resources can be partitioned into three categories: resources such as fossil fuels and minerals that will be depleted with consumption (exhaustible resources), resources such as forestry and fishing products that can be regenerated (renewable resources), and land and space resources. Animals only use renewable resources which are obtainable within a reachable area, thereby using only the amount required to exist. Furthermore, in ancient society, man used only the amount of resources required to exist from within their sphere of daily movement. In other words, by using resources from their surrounding natural world, mankind learned through many years of experience that excessive gathering would lead to a decrease in the amount of the resource gathered or the depletion of the resource. For renewable resources such as forests and fishing resources, sustainable use is possible if consumption does not exceed the speed of production, and there were conventions and rules in local communities to prevent the depletion of resources; mechanisms were in place to prevent a particular amount of collection of a particular resource at a level at which the resource could be renewed, thereby allowing for sustainable resource collection. In contrast, however, modern society has reached a point where more resources than are necessary are procured from all over the world, producing amounts of waste too large for the natural material circulation cycle to handle. As economic growth is prioritized, exhaustible mineral resources such as metals, rare metals, and fossil fuels including petroleum and coal are sought after all over the world, leading to a global resource problem. It is evident that the amount of these exhaustible resources decreases when they are consumed, thereby creating a definite need to enable their reuse. However, a drop in their quality upon reuse is unavoidable, and this therefore limits the uses of such reused resources.

In economics, goods that can be used inexhaustibly at no cost are defined as free goods. Water and air are free goods that come at no cost, and we therefore have a tendency to use them excessively. In addition, groundwater is pumped out and used in large amounts, forests and fishing resources are collected excessively, and carbon dioxide emissions continue to surpass environmental carrying capacity; all such practices can be identified as the root cause of modern resource and environmental problems.

9.3.2 Resource Consumption and Resource Recycling: Material Flow in Japan

A material flow shows the flow of material use from resource collection to disposal, thereby showing how much of a material (such as natural resources and recyclable resources) is used within the economic activity of a country and how much of these materials is consumed or disposed of. A material flow reveals a complete picture of the amount of natural resources being used, the amount of materials being reused, and the amount of waste being generated and thus provides potential ideas for policies that aim to build a recycling-oriented society.

Figure 9.1 shows the material flow in Japan for the fiscal year 2011 (Ministry of the Environment Homepage 2014). 1.57 billion tons of material was input, but the amount of recycled material only accounted for 0.238 billion tons of this amount, which is 15.2 % of the total material input. The amount consumed for energy and food is 0.574 billion tons, 36.6 % of the total; the amount of waste generated is 0.558 billion tons, 35.5 % of the total. The net increase of accumulated material in the figure shows the stocked amount of structures including vehicles, electronics, buildings, and roads, which are produced from the input of natural resources. The stocked amount is 0.511 billion tons, 32.5 % of the total material input. Therefore, the stock amount of materials surpassed 30 % of the total amount of material input. Materials in Japan are accumulating and are supplied both domestically and from around the world. These accumulated materials are a portion of materials that will become useless after a time lag and discarded as waste if not recycled. Such materials are also referred to as potential waste, and there is concern that they could become disaster waste if a natural disaster occurs, such as an earthquake or flood. During the Great Hanshin earthquake of 1992, 19.58 million tons of disaster

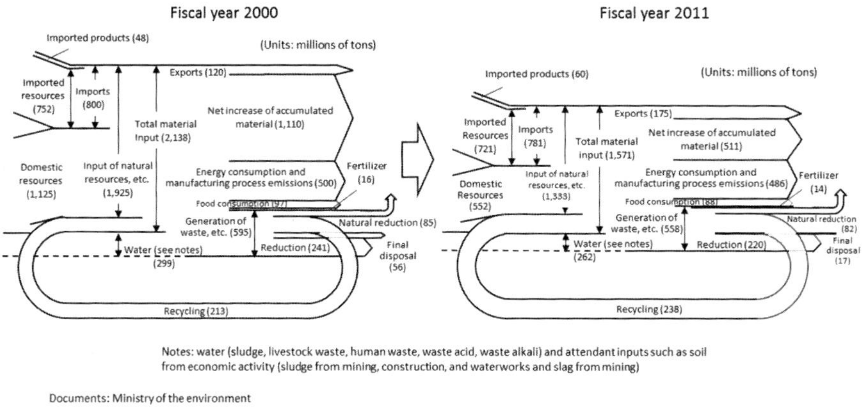

Fig. 9.1 Material flow in Japan

waste was generated. This amount was 8.5 times the yearly amount generated by Hyogo Prefecture, and such a disaster created a clear new awareness of the massive amount of materials stocked in structures such as buildings and homes (Hyogo Environment Create Center 1998).

9.4 Establishing a Recycling-Oriented Society to Solve the Resource Problem

9.4.1 Ways of Promoting the Formation of a Recycling-Oriented Society

In the latter half of the 1990s, despite difficulties with finding locations for the establishment of waste disposal facilities, there was a shift toward higher levels of waste generation, and the further advancement of recycling was called for. Additionally, due to factors such as the shortage of waste disposal facilities, the amount of illegally dumped waste was rising. With the background of such social conditions, the Basic Law for Establishing a Recycling-oriented Society was promulgated in June 2000, with fervent hopes of curbing resource consumption and creating a recycling-oriented society that would minimize the burden on the environment. This was to be accomplished by breaking away from an economic system of high production, high consumption, and high disposal and by recycling and effectively using materials from production through distribution, consumption, and disposal. The main points of the Basic Law for Establishing a Recycling-oriented Society are as follows:

1. Creating a society in which consumption of natural resources is controlled and the burden on the environment is as low as possible, by making a clear proposal

on the nature of the recycling-oriented society to be established and by ensuring that waste generation is curbed, recyclable resources are used in a cyclical way, and waste is disposed of properly.

2. Useful materials among the waste subject to the law are defined as "recyclable resources", and materials among the waste that are usable as resources are "recyclable resources" whether they have monetary value or not.
3. The law establishes a disposal priority ranking of (1) reduce, (2) reuse, (3) recycle, (4) thermal recycle (heat recovery), and (5) proper disposal.
4. The law clarifies the division of roles between the country, local governments, businesses, and citizens.
5. The law formulates the government's Basic Plan for Establishing a Recycling-oriented Society.
6. The law elucidates the nation's policies for establishing a recycling-oriented society.

9.4.2 Using Sustainable Environmental Technology in a Move Toward Recycling Resources

Technology incorporating the three elements below is considered ideal for turning large amounts of recyclable waste of varying quality into recyclable resources. This text introduces an example of this technology, which is referred to here as "sustainable environmental technology":

1. The technology contributes to building a recycling-oriented society from the standpoint of environmental burden, energy, and the economy.
2. It uses the laws of nature (natural phenomena) as its foundation.
3. It has a clear technological mechanism and is not easily altered by external factors, and it functions reliably and universally.

(a) Incineration bottom ash stabilization technology using carbonation treatment (Motohata et al. 2004)

Technology that utilizes the following properties found in incineration bottom ash is being proposed as technology for insolubilizing heavy metals such as lead, which is problematic in the effective use of municipal waste incineration bottom ash as a construction material. Moistened incineration bottom ash shows high alkalinity and thus absorbs carbon dioxide. During this process, carbonate and a heavy metal melted from the incineration bottom ash are formed, and elution of the heavy metal falls. This carbonate primarily forms membranes on the surface of the incineration bottom ash particles, thereby additionally suppressing elution physically. Furthermore, when carbonation progresses within the incineration bottom ash particles, they become delicate, and manifestation of individual particle strength can be expected. The carbon dioxide concentration in exhaust gas emitted from incineration plants is of the order of few percent, and the fact that the carbon dioxide in this exhaust gas is actively used in carbonation treatment also leads to the reduction of emissions into the air.

(b) Using dechlorination to create raw materials for cement from incineration ash (Shimaoka et al. 2011)

The high concentration of chlorine in incineration residues is problematic in relation to creating raw materials for cement from incineration residues. If the chlorine concentration of cement is high, rebars within concrete will rust, leading to deterioration of the concrete. However, insoluble chlorine that does not dissolve makes up 60–70 % of the total chlorine in incineration residues. Making incineration residues into raw material for cement requires simple, low-cost technology for use in desalinating the chlorine content of incineration residues. The effectiveness of dechlorination technology has been proven effective when the following phenomenon is applied. By mixing incineration bottom ash with organic matter such as biomass, the pH is decreased by the sulfate and the carbonate ions that accompany the breakdown of organic matter, and insoluble chlorine is easily decomposed. In this respect, solubilization is thus confirmed.

References

Closed System Disposal Site Development Association (2004) Closed system disposal sites for everyone. Ohmsha, Tokyo

(Foundation) Suginami Shoyo Memorial Foundation (1983) Tokyo's war on trash – Takaido Resident Records. Suginami Shoyo Memorial Foundation, Tokyo, pp 8–16

Hyogo Environment Create Center (1998) Records on disaster waste disposal. Hyogo Environment Create Center, Hyogo

Japan Waste Management Association (1996) Waste in Japan '96. Environmental Division, Water Service and Environment Department, Environmental Health Bureau, Ministry of Health and Welfare, Tokyo

Ministry of the Environment Homepage, Environment, Recycling-oriented Society, and Biological Diversity White Paper 2014 version (2015)

Mizoiri S (1988) A hundred years of garbage history – changes in disposal technology. Gakugei Shorin, Tokyo, p 393

Motohata T et al. (2004) Basic research concerning insolubilization of heavy metals from carbonation for effective use of incineration ash. Environ Eng Res Pap Collect 41:pp 459–467

Shigaki M (2000) Waste incineration technology (Revision vol 3). Ohmsha, Tokyo, pp 82–83

Shimaoka T et al. (2011) Dechlorination of municipal solid waste incineration residues for beneficial reuse as a resource for cement. J ASTM Int 8:10

Part III
Environmental Evaluation

Chapter 10
Concept of the Environment and Environmental Systems

Hirofumi Nakayama

Abstract In this chapter, we explain the basic concepts of the environment and environmental systems analysis. In Sect. 10.1, the concepts of environmental problems are explained using the environmental capacity relationship between the nature and scale of human activities. In Sect. 10.2, the basic concepts necessary for environmental protection are explained. Finally, in Sect. 10.3, the procedures and methods of environmental systems analysis are detailed in a systematic analysis of environmental problems focusing on the relationship between the natural environment and human activities.

Keywords Environmental capacity • Eco-efficiency • Basic concepts of environmental protection • Environmental systems analysis • Causal relationship diagram • Model analysis • Scenario analysis

10.1 Environmental Capacity of the Nature and Scale of Human Activities

The socioeconomic system developed thus far has extracted a large amount of natural resources, used these resources as raw materials to mass-produce intermediate and final products, and consumed and disposed of these products. In other words, it has been criticized as a system standing on mass production, mass consumption, and mass disposal. We enjoy convenient and rich lives by consuming a large amount of resources with excessive resource extraction, pollutant emissions generated during production, and unnecessary waste created in the consumption stage. These activities burden the natural environment, preventing material circulation in nature (Fig. 10.1).

The natural environment has the power to regenerate and restore itself even if it undergoes external changes. The regenerative and self-restorative capacity of nature is called environmental capacity. In the field of ecology, environmental capacity has

H. Nakayama (✉)
Department of Urban and Environmental Engineering, Kyushu University, Fukuoka, Japan
e-mail: nakayama@doc.kyushu-u.ac.jp

T. Shimaoka et al. (eds.), *Basic Studies in Environmental Knowledge, Technology, Evaluation, and Strategy*, DOI 10.1007/978-4-431-55819-4_10

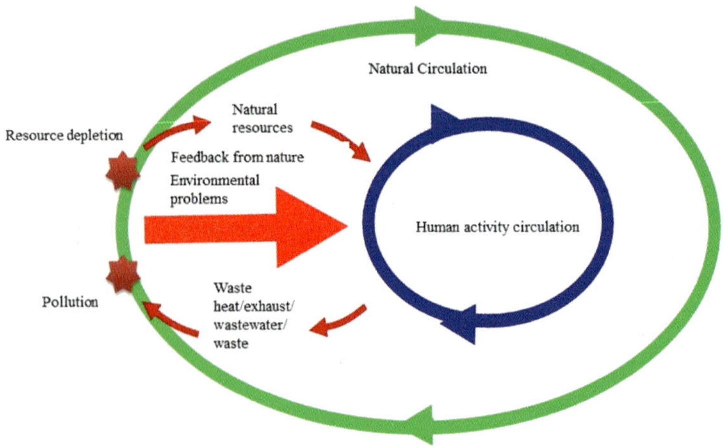

Fig. 10.1 Material circulations in nature, human activities, and environmental problems

been studied as the capacity of the natural environment for biomass and diversity. However, environmental capacity is limited. If a change occurs that surpasses the environmental capacity, the environment is damaged. Currently, mass extraction of resources from the natural environment and environmental burden of pollutants and waste from human activities are threatening human being survival. Therefore, resource consumption and pollutant emission from human activities must remain within nature's environmental capacity range. In addition, some technology measures to increase nature's environmental capacity have been considered.

Economist Herman E. Daly focused on transportation of materials between the natural environment and human society, for example, resource use and pollutant emission. He proposed the following three principles on sustainable rates (Herman 1990):

For a renewable resource—soil, water, forest, and fish—the sustainable rate of use can be no greater than the rate of regeneration.

For a nonrenewable resource—fossil fuel, high-grade mineral ore, and fossil groundwater—the sustainable rate of use can be no greater than the rate at which a renewable resource, used sustainably, can be substituted for it.

For a pollutant the sustainable rate of emission can be no greater than the rate at which the pollutant can be recycled, absorbed, or rendered harmless by the environment.

One method used to determine (1) the sustainable utilization rate of renewable resources is the maximum sustainable yield (MSY). It refers to the maximum sustainable harvest that does not reduce the level of resource stocks. The harvest of renewable biological resources (e.g., fish) must remain within the range of the net reproduction rate (i.e., the number of resource units increasing per hour) generated from a certain amount of resource stock. The maximum production possible while maintaining the stock is MSY (Ueta 1996).

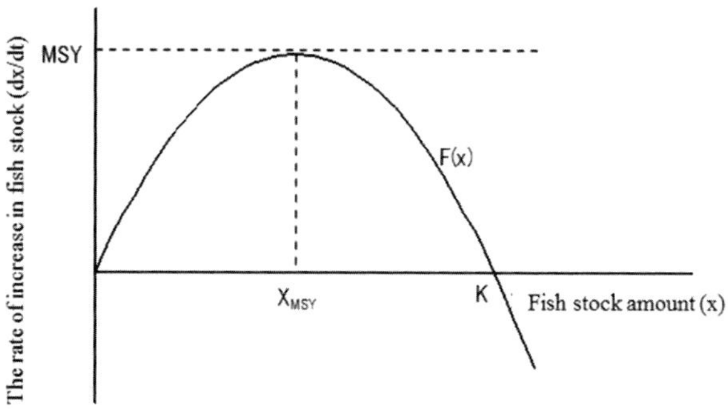

Fig. 10.2 A natural growth model of fish (*Source*: Ueta 1996)

Figure 10.2 shows a natural growth model of fish as a relationship between the rate of increase in the stock of fish and the existing stock amount. Considering a case where fish enters a new habitat with plenty of food, the rate of stock increase is significant. As the stock increases, the marginal rate of increase drops. When the stock amount reaches the environmental capacity K, the marginal rate of increase is 0. Then, the rate of increase peaks when the stock is one-half the environmental capacity K. If this level of stock is maintained and only the marginal quantity of fish is harvested, the maximum harvest can be maintained without exceeding the resource regenerative capacity.

This idea is utilized as the basic concept of using renewable resources in fishing, hunting, forestry, and agricultural production.

Next, for (2) the sustainable utilization rate of nonrenewable resources, we discuss the sustainable use of fossil fuels, such as oil. With a system that invests a portion of the profits from using oil in the development and promotion of technology for using alternative resources, alternative renewable energy can be continuously used even after the recoverable reserve is mined to exhaustion and the nonrenewable resource is depleted.

The use of renewable and nonrenewable resources must consider the resource regeneration rate and the rate of technology development for alternative resources. For the present and future population to enjoy a healthy and rich environment, this environment must be properly protected. The current generation cannot use an unlimited amount of resources to allow future generations to inherit these resources. Thus, fairness between generations—securing environmental resources for future generations—must be examined.

Finally, for (3) the sustainable emission rate of pollutants, there is a method of establishing environmental standards (that should be maintained to protect human health and secure the living environment) based on the relationship between this rate and environmental capacity. In other words, in this approach, environmental standards are set based on the balance between environmental load generated because of human activities and the purification capacity of nature, that is, environmental capacity. For example, if we can identify the environmental capacity of an aquatic environment, the balance between this value and pollution load measures such as biochemical oxygen demand (BOD) and the target water quality level based on use purposes can be used to designate environmental standards.

As discussed, to build a sustainable society where the environment is protected, the scale of human activities needs to be rationally controlled so that it stays within the range of environmental capacity. One method for doing this is an approach that increases the resources productivity and lowers the environmental load generated by human activities, titled Factor Four (von Weizsäcker et al. 1988). This approach has been argued by E. U. von Weizsäcker, Amory B. Lovins, and others concerning the target efficiency for resource utilization, where "factor" is the index for considering the degree of efficiency improvement. The population of developed countries and OECD member countries is only 20 % of the global population, but these countries use 80 % of global resources. From the perspective of fair access to resources and energy, this situation is unjust. To solve this, developed nations should reduce their resource consumption to one-fourth the current value. However, simply reducing resource consumption means abandoning quality of life, which is difficult to argue. Therefore, to reach this reduction, product performance and service amount can be doubled while necessary resource input for production is halved, quadrupling the efficiency (i.e., Factor Four). Factor 10 (Schmidt-Bleek 1997) was proposed by Friedrich Schmidt-Bleek of the Wuppertal Institute of Germany in 1991. This idea aims to achieve a factor of 10 by 2050 by reducing per capita resource consumption or environmental load emissions in developed countries while improving resource productivity.

The Rio Declaration stated that mankind plays a central role in sustainable development. According to this thinking, critical to environmental protection are various technologies for measures against pollution, technology related to equipment with high energy consumption efficiency, and biotechnology that scientifically controls and uses biological resources. Such an approach is termed anthropocentric environmental protection. On the other hand, there is an argument that the goals of sustainable development must be reviewed, and environmental protection should be prioritized over human living conditions. One of these ideas is deep ecology (Alan et al. 2001; Carolyn et al. 1994), which is a philosophy and movement initiated by the Norwegian philosopher Arne Næss. The word "deep" is based on the idea that environmental problems should be tackled from their roots. In contrast, the concept that aims to achieve sustainable development using technology development is termed "shallow ecology" (to avoid the word shallow, which has a negative connotation, "reformism" may be used). The philosophy of deep ecology accepts a certain level of necessity for shallow ecology, that is, a response to environmental problems with technological development. However, current environmental problems are considered to have surpassed the capabilities of shallow ecology. Considering the roots of environmental and social problems, values that pursue materialistic richness must be revised. Furthermore, instead of living standards and materialistic accumulation, quality of life and self-realization should be desired.

10.2 Basic Concept of Environmental Protection

1. Precautionary of prevention

   When the scale of environmental problems becomes large, mechanisms spanning from its phenomena to its effects become highly complex; thus, it becomes difficult to predict the seriousness of future potential problems. For global-scale environmental problems, it is difficult to discuss and determine the causal relationship between human activities and environmental effects. At this stage, because of insufficient scientific knowledge, whether some of the problems would actualize remains unclear. In such cases, people may not agree to sacrifice current benefits for potential future problems. However, some irreversible environmental problems cannot be recovered to current conditions once they actualize. In such cases, there is a stipulation that the lack of complete scientific certainty cannot be used to postpone measures with high cost/effect ratio to prevent environmental degradation. This principle of decision-making related to uncertain phenomena is called the precautionary principle (History et al. 2004). In Japan's Basic Environment Plan, "precautionary measures" is one of the four ideas for guidelines of environmental measures.
2. Source reduction

   The first pollution prevention measures restricted concentrations and amounts of pollutant emissions. Waste treatment measures have also been implemented to reduce and detoxify emitted waste. As such, measures that treat existing pollutants and waste are called end-of-pipe measures. In contrast, the design of products and production methods devised to reduce the creation of pollutants and waste are source reduction (History et al. 2004). In 1991 in Japan, policies on limiting emissions and recycling were introduced at a legal level for the first time: one law was enacted to promote the use of recycled resources and one regarding waste treatment and cleaning was amended. Later, limiting emissions were prioritized with the 1994 First Basic Environment Plan and the 2000 Basic Act for Establishing a Sound Material-Cycle Society. The latter Act stipulated that waste generation from raw materials and products must be controlled as much as possible. Furthermore, waste that still has some use was defined as a circulative resource, and principles of its use and disposal were stipulated. The Basic Act prioritized limiting emissions, reuse, recycle, heat recovery, and proper disposal, in that order.
3. Polluter pays principle and extended producer responsibility

   The polluter pays principle is a principle that states that the polluter is responsible for the costs of pollution prevention measures, environmental damage caused by pollution, pollution recovery, and relief for victims of pollutants mainly generated during the production stage of products. However, the polluter pays principle does not clearly state who is responsible for the cost of preventing and remedying environmental impacts generated during product use and disposal. In other words, it is unclear whether those who use the product or those who are responsible for its environmental impact after disposal are polluters. It is also unclear whether consumers who produce garbage or producers who make products that become garbage are responsible. This has led to the idea of extended producer responsibility, which is an approach of environmental policy that states that both production and post-consumption environmental load are the responsibility of the producer (History et al. 2004). Thus, product waste treatment responsibility shifts from the municipality to the producers, and incentives are awarded for environmental considerations during the product design stage. Here, there are both physical responsibilities, such as refuse disposal, and financial responsibilities, such as bearing treatment costs. In Japan, the Basic Act for Establishing a Sound Material-Cycle Society states that extended producer responsibility is the idea that "the producer must be responsible for their own products not only during the production and use stages but also when the product has become post-use waste." This is considered an extremely important perspective for establishing a recycling-oriented society for the following reasons: (1) the producer uses parts and materials repeatedly and continuously, so they may be able to reuse and recycle these materials; (2) producers have good knowledge of the product structure and properties and can easily separate and decompose its products; and (3) efforts to select materials and structures that contribute to reuse and recycling during the design stage are encouraged.

10.3 Methods of Environmental Systems Analysis

Environmental problems and environmental phenomena are systematically analyzed in environmental systems analysis (Environmental Problems & Kagaku 2009). Here, we describe typical methods used in this analysis. Systems analysis is a method for evaluating phenomena rising from interaction of multiple factors by subdividing the subject to clarify the characteristics of individual factors or by comprehensively examining the whole by integrating its factors. Generally, a system that exists for a certain purpose is composed of many interacting factors. In one system, multiple subsystems exist, and subsystems harmoniously interact with each other and exist as a coherent whole.

1. Extracting factors and understanding the causal relationship

   Typical procedures used in environmental systems analysis are shown in Fig. 10.2, and Table 10.1 summarizes the main methods used for each procedure. The first procedure includes processes (a) to (b), where a series of phenomena that lead to environmental problems are considered as a system. Here, a structure exists in which humans use the natural environment and receive amenity services, change, and affect the environment; this causes humans to experience the effects on the environment (Tomitaro Sueishi and Environmental Planning Research Group 1993; Committee on Environmental Systems Japan Society of Civil Engineers 1998). As seen, the essence of environmental systems analysis includes the extraction of various factors that affect the environment from human activities, the understanding of the causal relationship between those

Table 10.1 Methods used in environmental systems analysis

Procedure	Method
Factor identification, causal relationship summary	For example, brainstorming, the KJ method, the ISM method, the DEMATEL method, AHP, and structural equation modeling
Quantification of items such as environmental load, value, and damage	For example, LCA, MFA, monetary environmental valuation, environmental accounting, material flow cost accounting, and ecological footprints
Model description	For example, material balance model, SD model, and micro-/macroeconomic model (e.g., econometric model, I-O model, and CGE model)
Calculation and evaluation of the effects of introducing environmental items such as measures and technology	Forecasting and backcasting

factors, and structural modeling. For this purpose, methods such as brainstorming, the KJ method, the interpretive structural modeling (ISM) method, and the decision-making trial and evaluation laboratory (DEMATEL) method are used. These methods qualitatively express the structure through a diagram that summarizes the relationship between factors. Furthermore, structural equation modeling is used to quantitatively evaluate the compatibility of the generated relationship diagrams and the strength of the relationship between factors.

2. Qualitative evaluation of environmental load, value, and damage

   Next, a quantitative evaluation of environmental load from human activities and its impact on the environment is necessary. A representative method for the quantitative evaluation of environmental load is termed life cycle assessment (LCA), which evaluates environmental load created in a series of processes (life cycle) such as resource extraction, material manufacturing, production, distribution, sales, use, recycling, and disposal of products, services, or social capital. By evaluating the environmental load using the LCA, products and services with smaller environmental loads can be designed and developed. A typical LCA procedure consists of four stages: (1) setting goals and the evaluation range, (2) inventory analysis, (3) impact assessment, and (4) interpretation.

   To systematically understand the flow of materials and energy between the natural environment and economic activities as well as between economic entities (Moriguchi 2003), material flow accounting (MFA) is used. Under MFA, area and period are first sectioned. Then, the total input of materials into, flow of materials within, and total emission of materials outside of the area are estimated. In addition, using the data collected through the MFA, resource use efficiency is analyzed, termed material flow analysis. This method is used to identify waste of natural and other resources in an economy that cannot be identified in standard economic statistics. MFA is discussed in detail in another chapter.

A method that evaluates the size of environmental load and impacts considering its monetary value may be used. For example, the travel cost method, the hedonic method, and the contingent valuation method (CVM) are used for monetary evaluation of environmental values. Details of monetary evaluation of the environment are explained in another chapter. Under environmental accounting, environmental protection costs of business activities and the effects of these activities are quantitatively measured in monetary or physical units and then summarized and communicated. Environmental accounting focuses on wasted resources and energy loss generated during the production process and integrates material weight and environmental load emission. Then, material flow cost accounting is conducted to comprehensively evaluate the cost used for the loss in the production process, including the material, processing, and facility costs considered negative production costs. Integrated indexes that summarize multiple environmental loads and ones that combine economic and environmental indexes have also been developed.

3. Description of the model

   Next, a model is designed to quantitatively evaluate the connection between human activities and environmental impact. The appropriate model is chosen depending on the type of environment, goal, and policy measures. The forerunner of the environmental model is *The Limits to Growth* by the Club of Rome (Meadows 1972). Their model, the World3 model, considers various elements, such as population, industrial production, food production, and pollutant emissions, and their interaction structure is modeled using system dynamics (SD). Econometric models, input-output (I-O) models, and computational general equilibrium (CGE) models are examples of economic and material balance models that describe occurrence, movement, and alteration of environmental load. These models are sometimes used in combination with environmental load tables prepared using LCA and MFA (Nakamura 2007).

4. Scenario analysis (forecasting/backcasting)

   Finally, scenario analysis is conducted to predict the future environment (Fig. 10.3c). In scenario analysis, possible or ideal futures are tentatively visualized and quantitatively confirmed with model analysis. Here, there are two methods of preparing scenarios: forecasting and backcasting. Forecasting is a method that predicts the future deductively from current conditions and is also called trend analysis. In scenario analysis with forecasting, effects of multiple alternatives are evaluated for each scenario in the model (e.g., policy and technical progress), and it is determined whether the environmental index value reaches its final target level. Backcasting works backward from the goal. In this method, the ideal future and goals are determined, and necessary individual technology, systems, and organizations are considered. The "ideal future" set for backcasting is not singular, and multiple futures are visualized depending on their social acceptability and the concept of environmental protection. For example, a society with highly advanced technology (e.g., anthropocentric environmental protection) or one that values coexistence with nature may be visualized, and values of economic development and material wealth are

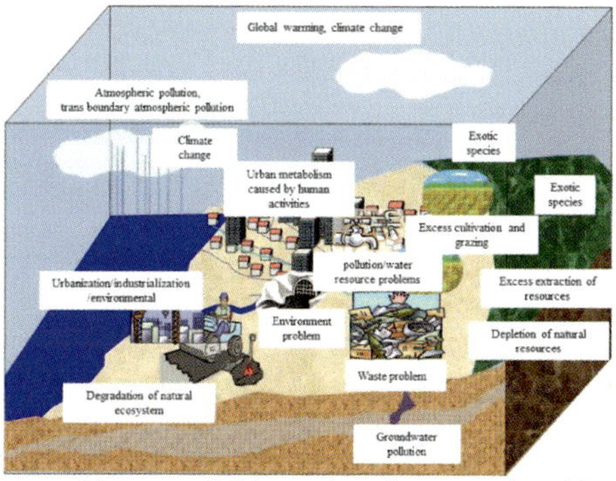

Human activities and environmental problems in society

↓

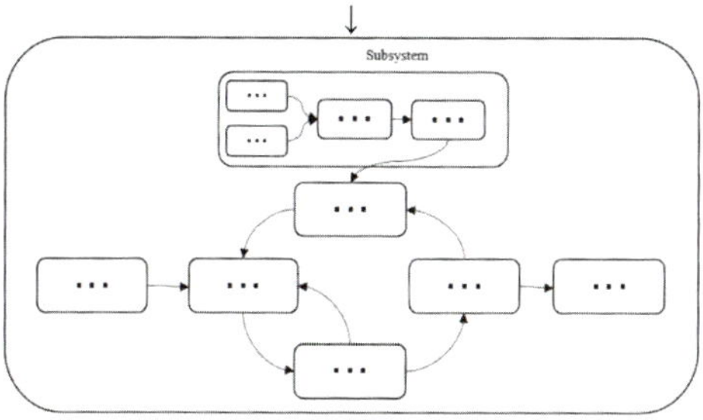

A model of the causal relationship between human activities and environmental problems

↓

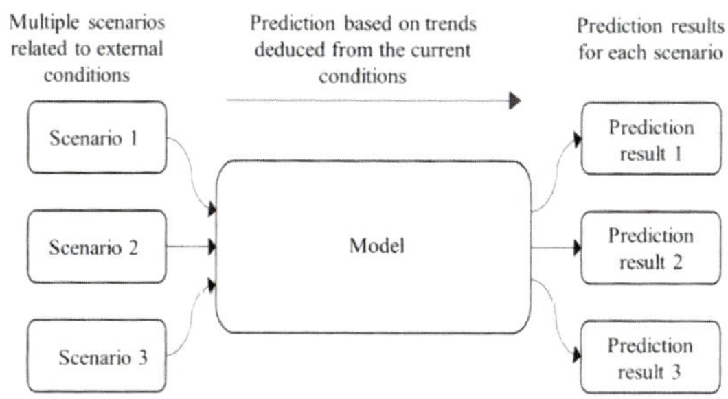

(c-1) scenario analysis (forecasting)

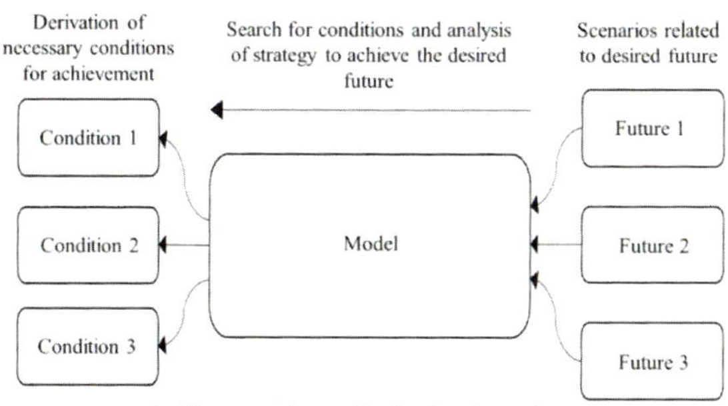

(c-2) scenario analysis (backcasting)

Fig. 10.3 Typical procedures of environmental systems analysis

revised. Critically, backcasting does not eliminate the potential that future societies may achieve by obsessing about ideals.

References

Committee on Environmental Systems Japan Society of Civil Engineers (ed) (1998): Environmental systems – concept and application method. Kyoritsu Shuppan Co., Ltd., Tokyo

Daly HE (1990) Toward some operational principles of sustainable development. Ecol Econ 2(1):1–6

Drengson A (2001) Deep ecology – environmental philosophy from the way of living (co-edited and trans: Inoue Y). Showado, Kyoto

Imura H (2009) Systematic approach to environmental problems. Kagaku Doujin, Kyoto

Kurasaka H (2004) Environmental policy – history, principle, and method of environmental policy. Shinzansha, Tokyo

Meadows DH (1972) Limits of growth – the club of Rome. Crisis of Humanity Report, Diamond Inc., Tokyo

Merchant C (1994) Radical Ecology – in search of comfortable world (trans: Kawamoto T, Sudo J, Mizutani H). Sangyo Books

Moriguchi Y (2003) Material flow data book – resource flow in the World Around Japan. National Institute for Environmental Studies, Tsukuba

Nakamura S (2007) Life cycle input output analysis. Waseda Institute of Political Economy. Research Monographs, Tokyo

Schmidt-Bleek F (1997) Factor 10 – eco-efficiency revolution (trans: Sasaki K). Springer-Verlag Tokyo

Tomitaro Sueishi and Environmental Planning Research Group (1993) Environmental planning theory – as a Foundation of Development and Protection of Environmental Resources. Morikita Publishing, Tokyo

Ueta K (1996) Environmental economics. Iwanami Shoten, Tokyo

von Weizsäcker EU, Lovins AB, Lovins LH (1988) Factor four: doubling wealth, halving resource use – a report to the club of Rome. Energy Conservation Center, Japan

Chapter 11
Socioeconomic Metabolism and Sustainability

Hiroki Tanikawa

Abstract This chapter focuses on the socioeconomic metabolism of material and energy flow and stock and sustainability. Although the flows of energy and materials should be balanced on a global scale, there are discrepancies at the individual and societal levels. Durable consumer goods, architecture, and society's foundational facilities support our livelihoods by providing various services by accumulating materials, thus also supporting our society with material stocks. However, among the materials necessary to support society, relatively large productions of concrete and ferrous materials are deeply linked with GHG emissions. Since they affect future global warming policies, it is necessary to design immediate measures for the sustainability of material stocks that support society.

Keywords Material stock and flow analysis • Energy flow • Sustainability • Hidden material flow • Urbanization

11.1 Earth's Material Balance

Our Earth maintains its planetary environment upon a delicate balance in space. The only external energy Earth receives is that which is radiated from the Sun to the Earth's atmosphere. Averaging over the whole planet, there is about 343 J per m^2 W of radiant energy from the Sun per second. In a year, this energy received amounts to $5*10^{24}$ J (5.5 yottajoules) for the entire Earth. Moreover, 30 % of this radiant energy is reflected back outside the Earth's atmosphere by clouds and the surface of the planet, while the remaining 70 % is transferred into energy such as heat, which temporarily stays on Earth (creating warmth) before being discharged back out to space as infrared radiation. As such, although there is a time gap, the outgoing energy from Earth is the same as the incoming energy (NASA 2004). Because of this, Earth does not have an extreme planetary environment like Mars or Venus, but instead maintains a stable, habitable environmental condition. The Earth's energy balance consists of output energy equivalent to the enormous input energy.

H. Tanikawa (✉)
Graduate School of Environmental Studies, Nagoya University, Nagoya, Japan
e-mail: tanikawa@nagoya-u.jp

T. Shimaoka et al. (eds.), *Basic Studies in Environmental Knowledge, Technology, Evaluation, and Strategy*, DOI 10.1007/978-4-431-55819-4_11

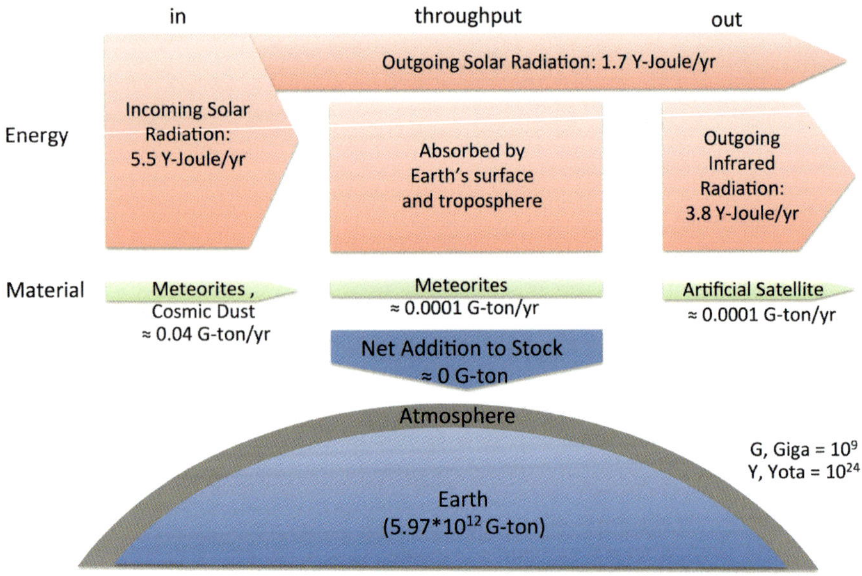

Fig. 11.1 Energy and material balance of the Earth (Estimated from NASA (2004))

As for the material balance on Earth, however, although there are meteorite collisions and rockets being launched, there is almost no change in the total mass of the Earth, as those changes are negligible compared to the planet's total mass. Therefore, the total material balance on Earth, unlike its energy balance, does not consist of much input or output; rather, it stays constant. The energy and material balances on a global scale, though different in shape, both maintain equilibrium (at least for short periods). In an extreme case, it would be very difficult for humans to restore the delicately maintained energy balance between the outside and inside of Earth if the scales were ever tipped. On the other hand, larger space systems, such as the solar system, would not be largely affected due to the little change in the material balance between Earth and outer space. It is crucial to recognize the direct connection between human activities and the sustainability of life on Earth (Fig. 11.1).

11.2 Energy and Material Balance of Human Society

The energy and material balances on Earth depend largely on boundary settings. Within a single country, for example, the energy and material balances change greatly depending on geography, abundance (or lack) of resources, economic conditions, level of societal maturity, and development of technologies. Energy flow analysis and material flow analysis generally target entire countries and are indispensable tools for reflecting the country's condition and analyzing its contributions to future resource strategies and the Earth's overall environment. When looking at an individual company, it is possible to measure its resource and energy consumption by the products and services it provides, as well as its emissions, such as waste materials and CO_2. The company's economic efficiency, therefore, can also be considered its performance index. In either case, it is important to quantify, in chronological order, the materials and energy consumed and discharged (or produced) in the subject region (country, municipality, corporation, or individual).

For example, Brunner and Rechberger (2004) describe the change from the past to the present in the material balance per capita as follows. Hunter-gatherers in the prehistoric age would attain 6 tons per capita of food and drinking water in a year. Of those 6 tons of material input, 5.1 tons would be digested, used as energy, and discharged into the air through respiration. The remaining 0.9 tons would become excretions and waste. Therefore, their lives were based on the 6 tons of annual resource consumption. Today, on the other hand, the annual resource, energy, and water inputs per capita are 86 tons (Brunner and Rechberger 2004). Through various activities, 3 tons are accumulated (i.e., stationary, stable stock such as houses and cars), 19 tons discharged to the air, 61 tons discharged into the sewer, and 3 tons become other solid wastes. The material balance of the humans in the prehistoric age was purely for their survival. Taking into consideration their shorter life spans and fewer cultural aspects, the difference in material balance between past and present clearly supports the suggestion that modern lives require more abundant energy and material stocks.

For the material and energy flows required today, Sankey diagrams are often used, which express flow rates by the thickness of the arrows. Figures 11.2a and 11.2b, made by the International Energy Agency (IEA), shows the changes in global energy flows using a Sankey diagram. It can be observed that, over 40 years, the global energy flow doubled, in tonnes of oil equivalent (TOE). Needless to say, this figure only shows the energy flow from its source to consumption and not the waste side, which makes it an incomplete representation of the balance; however, since the waste part of the energy flow mainly consists of discharge to the environment through heat and CO_2 along with the workload, there is an approximate balance between the input and the output. On the other hand, Figs. 11.3a and 11.3b shows the change in global material flow in terms of weights from the perspective of material cycles (NIES CGER Report 2006; Moriguchi and Hashimoto 2006). When comparing 1983 and 2003, the global flows of fossil fuels, agricultural resources, forestry, fisheries, and metal resources increased, particularly in East and South Asia. Fossil fuels are expressed with the thickest arrow, indicating its magnitude as a material flow vector. As stated above, fossil fuels become the source of greenhouse gases such as CO_2 when burned (i.e., at the end of the arrow). The Center for Global Environmental Research of the National Institute for Environmental Studies (NIES CGER) report also notes that the CO_2 emissions through the usage of fossil fuels are the largest human waste.

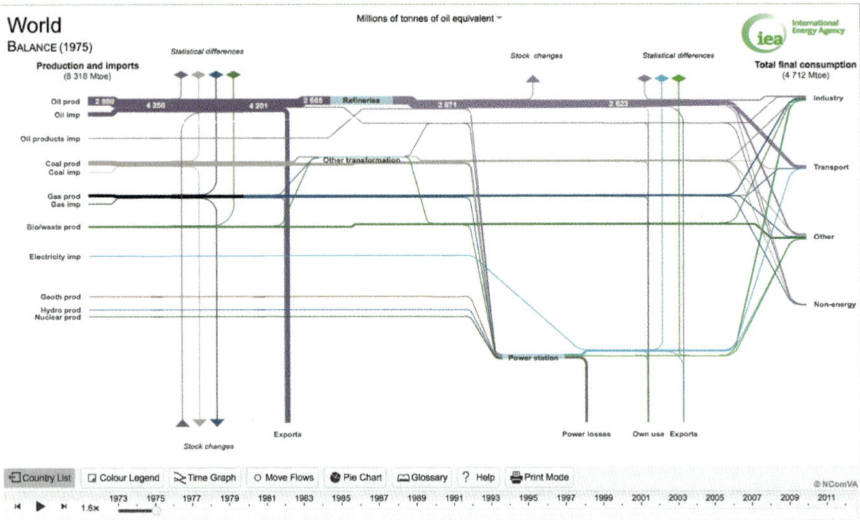

Fig. 11.2a IEA energy balance Sankey diagram, 1973 World Energy (*Source*: http://www.iea.org/)

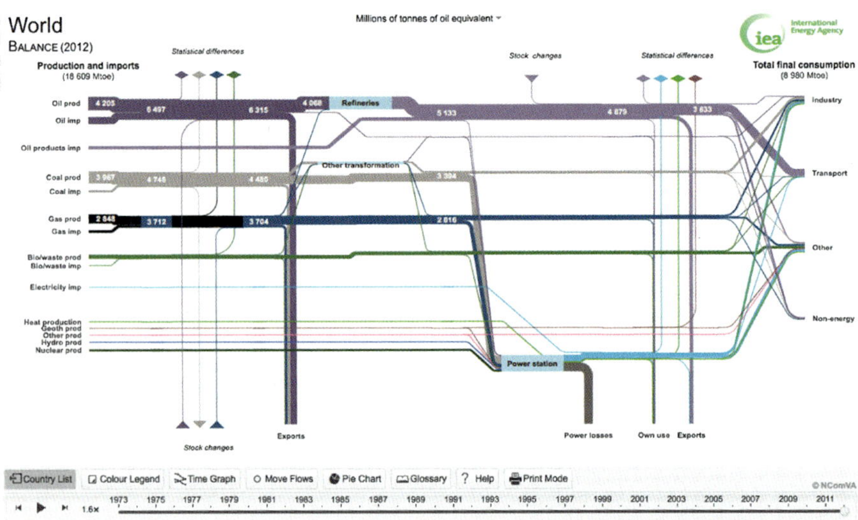

Fig. 11.2b IEA energy Balance Sankey diagram, 2012 World Energy (*Source*: http://www.iea.org/)

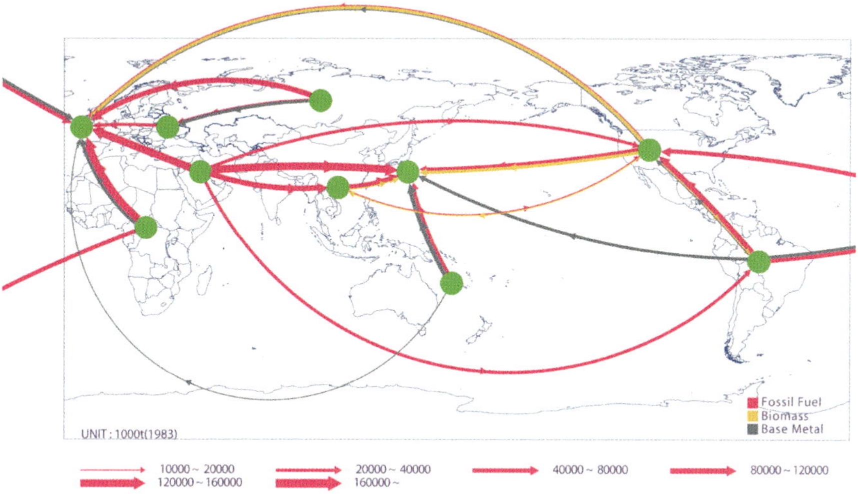

Fig. 11.3a Trade flow of the main resources in the World, 1983 (*Source*: NIES CGER Report 2006, Moriguchi and Hashimoto 2006)

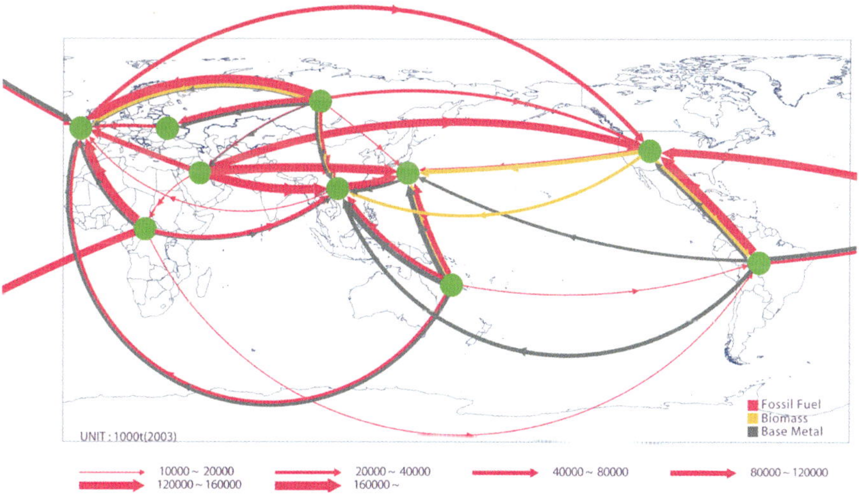

Fig. 11.3b Trade flow of the main resources in the World, 2003 (*Source*: NIES CGER Report 2006, Moriguchi and Hashimoto 2006)

11.3 Abundance and Material Stocks

When considering the amount of resources and energy required to enrich human lives, the difference between prehistoric and modern material balances becomes evident. While humans required only 6 tons of resources per capita per year in the

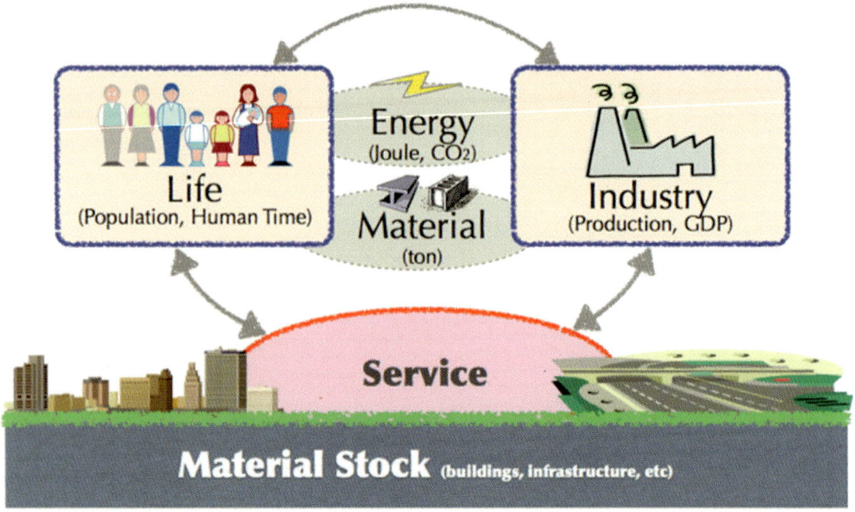

Fig. 11.4 Material stock that supports society

prehistoric age, people in the modern age require 86 tons of resources per capita per year. In other words, there is a difference of 80 tons of necessary resources for modern, enriched lives. Besides the large difference in the material flow, another significant difference is the material stocks. While the prehistoric age saw little preservation of material stocks, currently, there is an annual accumulation of 3 tons and a total of 260 tons of stock preservation per capita. The material stock consists of durable consumer goods such as houses, cars, and furniture, which stay in place and offer services to humans. Familiar material stocks include furniture, electrical appliances, cars, and even infrastructures such as roads, waterworks, and sewers. In other words, as long as such materials persist, they not only support human lives but also provide various benefits such as convenience and safety, which did not exist in the prehistoric age (Fig. 11.4).

11.4 Material Flow, Stock, and Global Warming

In order to attain a state of abundance, a certain amount of material stock is necessary for any civilization or lifestyle. In order to achieve this, resources are needed to make stocks (e.g., wood and plastics for furniture and metal resources such as iron and aluminum to produce cars). The material stocks that are heaviest in urban areas include buildings such as houses and societal infrastructures, which require resources such as soil and stone, concrete, and steel.

Cement, a material necessary to make concrete, requires limestone, clay, and silica as its ingredients and coal as a heat source. As shown in Fig. 11.5, the basic process involves the extraction of CO_2 through calcination of $CaCO_3$, the

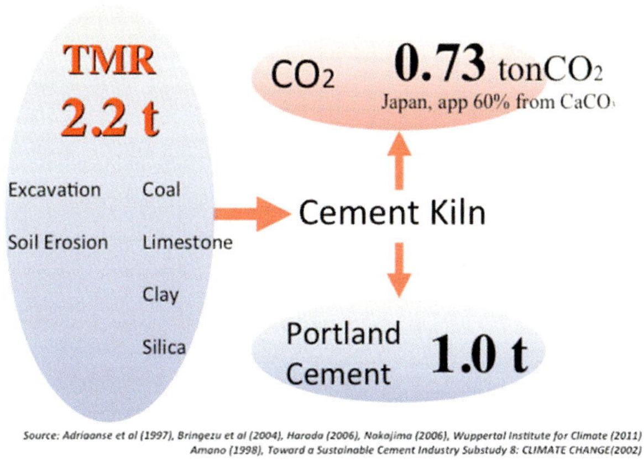

Fig. 11.5 CO_2 emissions and materials required for cement production

main component of limestone, and production of CaO, the main ingredient of cement. Here, clinker (i.e., lumps or nodules) is created by mixing clay and silica, which requires a heat source that produces a temperature higher than 1000 °C (i.e., thermolysis). As such, coals and waste materials are burned. Let us look at the process of cement production from the viewpoint of material flow. First, in order to secure the necessary ingredients, the excavation and selection of raw materials such as limestone and coal are needed. In this process, although rock wastes occur through excavation of topsoil and ore dressing at the mine, these factors are not included in general statistics because they do not accompany economic activities. Such material flows that are due to human activities done outside economic activities are called hidden material flows (HMF, or ecological rucksacks). Moreover, the material input directly required for cement production is called direct material input (DMI), and the sum of HMF and DMI is called the total material requirement (TMR, or total associate material input). Here, TMR signifies the total material input used for cement production. Figure 11.5 shows the TMR for 1 ton of cement, based on data from Japan. In order to produce 1 ton of cement, the TMR is 2.2 tons, indicating the magnitude of environmental effect associated with cement use for generating stock. At the same time, the CO_2 emissions associated with producing 1 ton of cement are 0.73 tons, contributing to global warming, as does the TMR. These quantities vary depending on the production method and the extent of using waste materials.

In the same way, Fig. 11.6 indicates the TMR and the CO_2 emissions associated with steel production. Despite the variation due to different production methods, the TMR and the CO_2 emissions associated with producing 1 ton of steel are approximately 8.0 tons and 2.0 tons, respectively. Countries with a higher level of production technologies have lower TMR and CO_2 emissions for producing steel.

The CO_2 emissions from cement production, as explained above, consist of burning fossil fuels and decomposing limestone through thermolysis. While this contribution to global warming could be lessened by technological improvements for better efficiency and alternative fuels, the latter is difficult to reduce unless limestone is replaced by an alternative raw material. Although it is possible to use

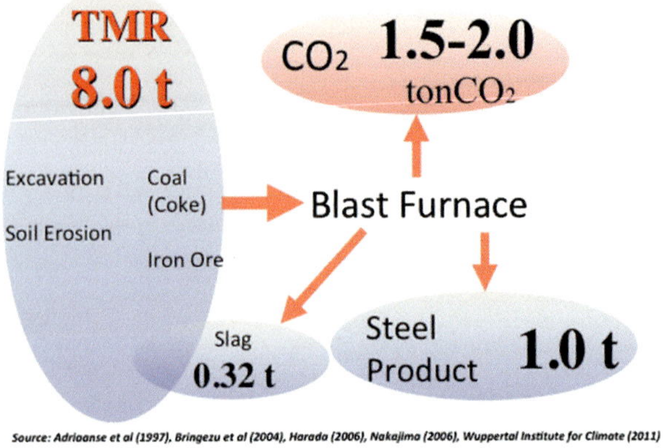

Fig. 11.6 CO_2 emissions and materials required for steel production

fly ash and steel slag as a portion of the alternative resources, it is difficult to replace the entire supply. According to the US Geological Survey (USGS, 2014), the global production of cement was approximately 3.8 billion tons in 2012, which was twice as large as in 2002, mainly due to the rise in production in China. The Intergovernmental Panel on Climate Change (IPCC) reported the CO_2 emissions from cement production were 0.78 t-CO_2/t-cement in 2010 on the Fifth Assessment Report (IPCC AR5, 2014). Multiplying this rate provides an estimate of the total global emissions of CO_2 associated with cement production, which is 3 billion tons. Since the total global CO_2 emissions in 2012 were reported to be 31.7 billion tons, the emissions from cement production comprised 9 % of the total.

In addition, steel production requires a large amount of thermal energy for the reduction of iron ores. Total global steel production in 2012 was 1.56 billion tons (Japan Iron and Steel Federation) with the CO_2 emission rate reported at 2.2 t-CO_2/t-crude steel (IPCC AR5, 2014), making the total CO_2 emissions about 3.4 billion tons for steel production. Its contribution to the total global CO_2 emission is approximately 11 % (Fig. 11.7). As for new technologies to reduce CO_2 emission, hydrogen reduction of iron ores will replace the conventional method, which uses coke oven gas. The industry is aiming to introduce this technology by 2030, with a goal of reducing CO_2 emissions by about 30 % at the end of this period; however, timely measures are expected since there is an emerging demand for steel and intense cost competition.

As cement and steel are indispensable for foundational facilities that support human activities, global demand is anticipated to rise even more in the future. Today, global CO_2 emissions associated with cement and steel production make up about 20 % of all emissions, and this figure is anticipated to rise due to increasing demand likely outgrowing the improvement in production efficiency. For example, if the goal was to restore the CO_2 emission level to what it was in 1990 (20.9 billion t-CO_2), the contributions to CO_2 emissions by cement and steel production will continue to rise, indicating that future material stocks (particularly buildings and societal and foundational facilities) will be important factors for global warming provisions.

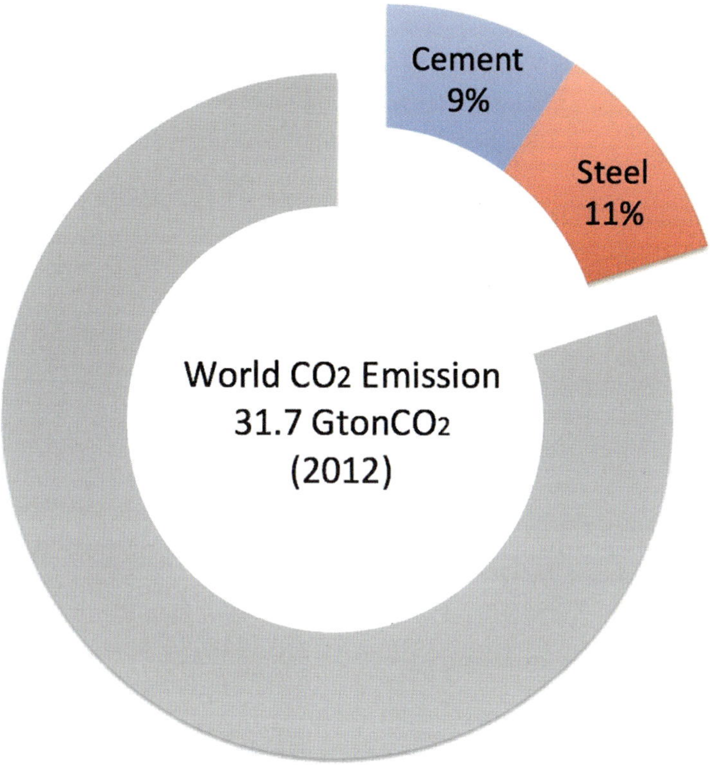

Fig. 11.7 Global CO_2 emissions associated with cement and steel production (2012)

References

Brunner PH, Rechberger H (2004) Practical handbook of material flow analysis. Lewis Publishers, Boca Raton

International Energy Agency (2015) IEA Sankey Diagram. http://www.iea.org/Sankey/

IPCC (2014) Fifth assessment report on climate change: WGIII mitigation of climate change. http://www.ipcc.ch/

Moriguchi Y, Hashimoto S (2006) Material flow data book, world resource flows around Japan. CGER Report. National Institute of Environmental Studies, Japan

NASA (2004) The Earth's Energy Budget http://www.nasa.gov/audience/forstudents/5-8/features/F_The_Role_of_Clouds.html

NIES CGER Report (2006) Material flow data book, world resource flows and around Japan. http://www.cger.nies.go.jp/publications/report/d040/D040.html

US Geological Survey (2014) Mineral commodity summaries

Chapter 12
Environmental Problems in Economics

Hideo Koide

Abstract Economics is a specialized field in social science that studies the fundamental problems of how best to allocate limited resources between economic agents to achieve optimal results. Typically, the mainstream economics subjects offered by a university economics faculty are macroeconomics and microeconomics. However, because of recent academic developments, students can now also read environmental economics and resource economics as part of the mainstream methodology. Section 12.1 examines the relationship between the environment and economics using the figures of Herman E. Daly, who originated ecological economics. Here, the focus is on those points not considered in so-called mainstream economics. Then, Sect. 12.2 provides a chronological discussion of the optimal usage of exhaustible resources as the introductory theme to resource economics. Here, "Hotelling's Rule," which is introduced as a propositional statement, indicates that the price necessarily needs to increase as the quantity of a resource decreases. Lastly, Sect. 12.3 investigates whether Hotelling's Rule is evident in data on the quantity of resources and prices. This section also discusses economics issues that arise from the analysis of environmental problems.

Keywords Environmental economics • Resource economics • Exhaustible resource • Hotelling's Rule

12.1 Economics and Environment

12.1.1 What Is Economics?

Economics is a specialized field in social science that studies the fundamental issues of how best to allocate limited resources between economic agents to achieve optimal results. Limited resources are, by their very nature, scarce, and in our daily lives, we often think of ways to use them efficiently, even if we do not know much about economics. For example, "how can I spend the 24 hours in my day?" "what

H. Koide (✉)
Seinan Gakuin University, Fukuoka, Japan
e-mail: koide@seinan-gu.ac.jp

T. Shimaoka et al. (eds.), *Basic Studies in Environmental Knowledge, Technology, Evaluation, and Strategy*, DOI 10.1007/978-4-431-55819-4_12

should I buy with my monthly allowance?" or "how can I allocate the tasks in my job when I have only two subordinates?"

The term "resource" includes natural resources, such as energy resources and mineral resources, but also includes abstract resources, such as goods (assets, services), money (funds), people (labor), and time.

In economics, the results of allocating these resources are measured in terms of a benefit or a net benefit, which is the amount remaining after subtracting the costs incurred. Benefits and costs are measured in monetary units, such as the yen or dollar. When the net benefit is maximized after allocating our resources, this is viewed as a favorable allocation in economics terms.

We can also use utility and profit to measure the resource allocation results, depending on the subject of the analysis. Utility measures the amount of satisfaction gained from using the goods and is a concept peculiar to economics. Profit is a measure of the amount gained from selling goods after we have subtracted our costs of producing the goods.

University economics faculties all offer subjects such as macroeconomics and microeconomics, and students read economics based on these ideas. Then, they read the applied fields of international economics (goods), financial theory (money), labor economics (people, time), environmental economics, and resource economics and study how economics can explain real-world problems.

12.1.2 The Relationship Between the Environment and Economics

There are other fields such as political economic science, economic history, history of economic thought, and theory of comparative economic systems that are not based on mathematical principles. Others focus on actual conditions and systems rather than mathematical principles. Many universities also offer Marxian economics, which they present as the principles of economics.

Researchers in such fields refer to the typical framework of macroeconomics and microeconomics as mainstream economics, or neoclassical economics, pointing out weaknesses in mainstream economics theory and sometimes presenting alternative ideas. This movement ultimately went beyond the academic society of economics, creating a unique field of arts and science, integrated into ecological economics. Ecological economics combines material balance, the law of increasing entropy in thermodynamics, and ecological systems.

Herman E. Daly, who originated ecological economics, indicated several points that mainstream economics misses, as shown in Figs. 12.1 and 12.2. First, our economy is included in the ecological system (environment). The economy takes in materials and energy from the ecological system, sometimes recycles them, and finally returns materials and energy, which have become useless. In addition, the

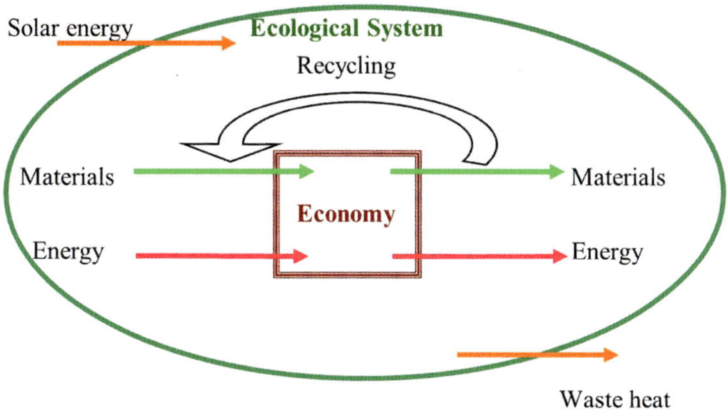

Fig. 12.1 Relationship between the environment and the economy (1): empty world (Source: Daly (2005) and Stiglitz and Walsh (2012))

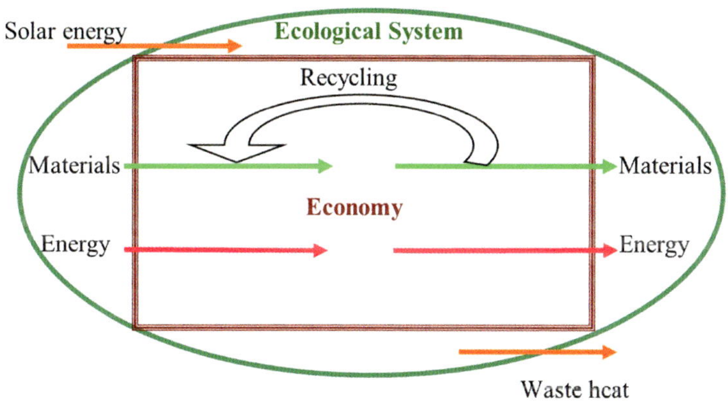

Fig. 12.2 Relationship between the environment and the economy (2): full world (Source: Herman E. Daly (translated by Isao Nitta, Shinobu Kuramoto and Masayuki Omori), "Beyond Growth: The Economics of Sustainable Development", Misuzu Shobo P. 69 (2005))

ecological system takes in solar energy from outside the system and emits waste heat.

The difference between Figs. 12.1 and 12.2 is the scale of the economy included in the ecological system. Figure 12.1 shows the economy in an empty world, as compared to the ecological system. Figure 12.2 shows the economy in a full world, almost bursting the ecological system.

The "optimal scale" of economy is the desirable maximum scale of the economy in relation to the environment. Surprisingly, mainstream economics cannot determine this optimal scale, because its definition of an economy does not include the natural environment.

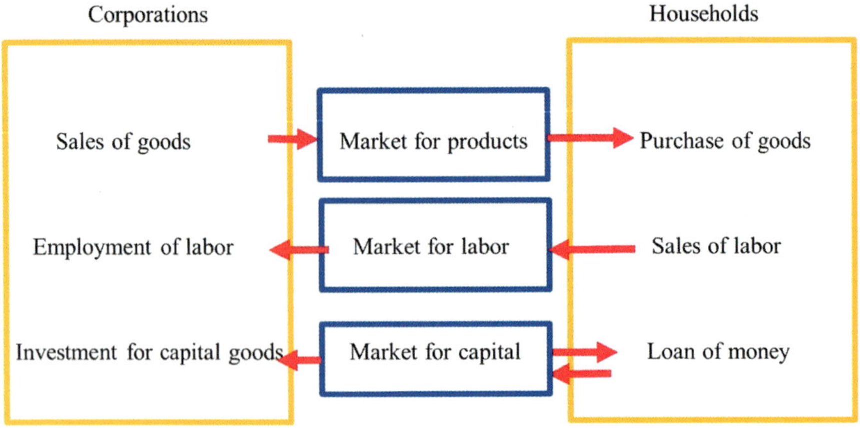

Fig. 12.3 Economic agents and the market as presented in economics textbooks (Source: Joseph E. Stiglitz and Carl E. Walsh (translated by Shiro Yabushita, et al.) (2012) Economics 4th ed. Toyo Keizai Inc. p. 22)

Note that the concept of an equilibrium national income (or equilibrium output) is taught at an early stage in macroeconomics. This refers to the scale of the national income in which the demand for goods and services is matched by their supply in the overall economy. However, this concept does not consider whether the economy is optimal in terms of the environment.

Figure 12.3 shows the flow and market of goods, people, and money as typically presented in (mainstream) economics textbooks, according to Joseph E. Stiglitz et al. The diagram shows two economic agents, which can be corporations or households. If governments and foreign countries are included as economic agents, the flow becomes more complicated.

The "Market for Products" in Fig. 12.3 do not explain how the goods appear, where the goods go after being used, or, if the economy is an independent system, whether the materials/energy to be used already exist and nothing is discarded after use. In other words, it is a system of perfect recycling. Needless to say, this is not a realistic economy.

A resource constraint is often assumed in mainstream economics. This means that the total amount of labor and capital stock in an economy is set. The concept of a resource constraint is also used for other "resources," such as limited natural resources or a limited amount of space that can be used for landfills.

12.1.3 Weaknesses of Mainstream Economics

Keeping in mind the connection between the environment and the economy, as indicated by Daly, we can identify three "weaknesses" in the framework of mainstream economics.

Question (1): How do the goods appear? Answer (1): The goods are produced by the combination of labor and capital. If we denote labor as L, capital as K, and production as Y, we have the production function $Y = f(L, K)$. Additional question (1): Should we not also include the input of natural resources and energy?

Question (2): Where do the goods go after being used? Answer (2): This question is not considered because economics deals with stages of production and consumption only. Even if waste is generated, it disappears at no cost. Additional question (2): Should economics not also include the problems of waste treatment and recycling?

Question (3): Is unlimited economic development possible? Answer (3): Unless there is some restriction to restrict growth, quantitative growth is possible. Additional question (3): Based on Figs. 12.1 and 12.2, how does an economy expand beyond the capacity of the environment?

The additional questions raised here are not typically considered by students of macroeconomics or microeconomics. They only become aware of these points when studying environmental economics or resource economics, which complement mainstream economics, or when studying an applied field in economics.

Note that the single term "environmental economics" can include mainstream economics, as well as subjects not based on mathematical principles. This point is peculiar to social science. Different methods are used to analyze one phenomenon.

An introductory problem in resource economics, introduced in this chapter, is how to apply economics to solving the problem of resource exhaustion. A further criticism of mainstream economics is its assumption that economic agents necessarily behave rationally (Söderbaum 2010). However, an introductory argument using mathematical principles is based on scientific methods and has significant merit.

Note that the textbooks on ecological economics by Herman E. Daly et al. provide a detailed explanation of mainstream economics, while also describing the importance of understanding (Figs. 12.1 and 12.2) (Daly and Farley 2004).

12.2 Optimal Use of Exhaustible Resources

12.2.1 An Increase in the Price of Resources

Figure 12.4 shows the change in the optimal price of an exhaustible resource derived from the theory of economics. Exhaustible resources are those that decrease in quantity as they are used, for example, fossil fuels such as coal and oil.

Here, t, on the horizontal axis, represents the period (e.g., a year), and the price of the resource in period t (or after t years) is p_t, as shown on the vertical axis. Furthermore, suppose the price p_0 at the initial point is 1. The price of the resource

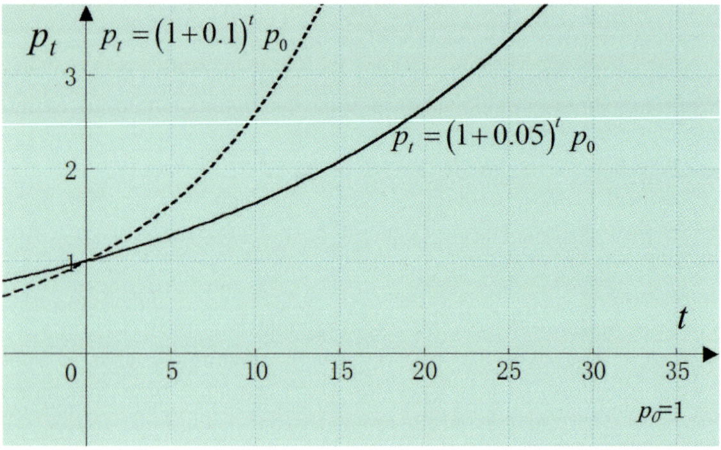

Fig. 12.4 Changes in the optimal price of exhaustible resources (Source: Author. t denotes the period (year), p_t denotes the price of the resource in period t, and p_0 is the price of the resource at the initial point)

is represented by the exponential function $p_t = (1 + r)^t p_0 = (1 + r)^t$, which passes through the vertical axis at 1. Here, r indicates the interest rate.

Two curves are shown in Fig. 12.4. The solid line is the price of the resource when the interest rate is 5 % ($r = 0.05$), and the dotted line is the price when the interest rate is 10 % ($r = 0.1$). A higher interest rate means the rate of the change in price increases more steeply.

Unlike a renewable resource, such as fishery resources and forest resources, an exhaustible resource cannot increase in quantity. Therefore, the amount available at the initial point reduces with use.

When the market mechanism of the economy functions properly, the price of a good is determined by the point at which supply and demand become equal. However, the amount that can be supplied from an exhaustible resource decreases over time. In other words, resources become scarce. As a result, the price of the resource will gradually increase over time. As the resource nears the point of exhaustion, the market price is expected to reach a very high level.

12.2.2 Hotelling's Rule

As mentioned earlier, the concept of a resource includes goods, money, labor, and time within mainstream economics. In contrast, resource economics focuses on how materials and energy are input from the environment into the economy and how to use these resources, including recycling, as well as prices and economic results. Natural resources have certain characteristics. First, the optimal use of an exhaustible resource is explained.

Before the field of resource economics existed, Harold Hotelling, an economist, spoke of the optimal use of an exhaustible resource over time (Hotelling 1931). His conclusion later came to be known as "Hotelling's Rule" and is the first propositional statement that students of resource economics learn.

There are numerous ways to use Hotelling's Rule. Suppose there is a form of economy where a resource of amount X is used up in the current year ($t=0$) and next year ($t=1$). This is the simplest form. The amounts of the resource used in each year are x_0, x_1, respectively. Since the resource will be used up after 2 years, we assume that $X = x_0 + x_1$.

Next, the utility (degree of satisfaction) gained by economic agents by using resource x is assumed to be $u(x)$. As x increases, u increases accordingly. However, the degree of increase will decline gradually, which is shown in differential calculus as $u'(x) > 0, u''(x) < 0$. Here, u' is called the marginal utility.

Note that the term marginal is used to mean "additional" in the context of economics. It does not mean "limit."

We use another method in economics, called discounting, when we add values at different times. When an interest rate (or discount) is assumed as $r > 0$, the sum of the utility gained in this year and the next is $u(x_0) + u(x_1)/(1+r)$.

As an aside, consider the following example. Suppose you deposit 100 yen at a bank this year. Interest is added on this principal in the next year. Suppose the interest rate is at 5 %. Then, the deposit at the beginning of the next year is $(1+0.05) \times 100$ yen $=105$ yen. We refer to this calculation of interest as compound interest.

Discounting is the reverse of the interest calculation. Therefore, using the same interest rate, the value of 100 yen next year, but assessed this year, is 100 yen $\div (1+0.05) \approx 95.2$ yen.

Now let's return to utility function, $u(x_0) + u(x_1)/(1+r)$, which we call the "discounted present utility" of an economic agent in the current year. If we rewrite the formula for resource exhaustion as $x_1 = X - x_0$, then substitute the RHS into the utility function, we have $u(x_0) + u(X - x_0)/(1+r)$. We now have the discounted present utility expressed as a function of x_0 only.

In economics, the operative variable is deemed to be zero after differentiation or partial differentiation in order to maximize the objective function. We can maximize the discounted present value by differentiating x_0 to be zero, giving us $u'(x_1) = (1+r)u'(x_0)$.

Furthermore, the prices of the resource this year and next year are assumed to be p_0, p_1, respectively. The economic agents use resources where the price is equal to the marginal utility u' to maximize their utility each year.

Therefore, we have $p_1 = (1+r)p_0$. This formula shows that the price of the resource next year is equal to the price of the resource this year multiplied by (1 + interest rate). This is the simplest form of Hotelling's Rule, namely, when there are only two periods.

The connection between the prices of the resource in the two periods is easily extended to T years (Conrad 1999). Note that the result is shown here, but not the calculation. The price of the exhaustible resource in t years is equal to the initial price p_0 multiplied by (1 + interest rate) t times. In other words, $p_t = (1+r)^t p_0$. The two curves in Fig. 12.4 show the case where p_0 is 1.

12.2.3 Examples

So far, we have discussed abstract theory. Here, we show examples of how to calculate the use and price of resources when they are used up in 2 years, assuming a specific utility function.

First, the utility gained by the economic agents from using the resource is assumed as $u(x) = x^{\alpha}$. We also assume that $0 < \alpha < 1$. Differentiating once and twice with respect to x gives $u'(x) = \alpha x^{\alpha-1} > 0, u''(x) = \alpha(\alpha-1)x^{\alpha-2} < 0$, respectively, for the range of α, satisfying the condition of the utility function.

When the resources are used up in 2 years, the discounted present utility of economic agents in the current year is $x_0^{\alpha} + x_1^{\alpha}/(1+r)$. Then, recall that $x_1 = X - x_0$. Substituting this into the discounted utility function gives the objective function $x_0^{\alpha} + (X - x_0)^{\alpha}/(1+r)$.

This objective function is a function of x_0. We find the solution to $x_0 = \frac{(1+r)^{1/(1-\alpha)}}{1+(1+r)^{1/(1-\alpha)}}X$ by differentiating it to be zero and adjusting. Next, the value is assigned to $x_1 = X - x_0$ to give us $x_1 = \frac{1}{1+(1+r)^{1/(1-\alpha)}}X$.

Here, we assume the following values, $\alpha = 0.5$, $r = 0.05$, $X = 10$, and then calculate the answer using a scientific electronic calculator or computer software. The amount of each resource used and their prices in each year are $x_0 = 5.244$, $x_1 = 4.756$, $p_0 = 0.218$, $p_1 = 0.229$, respectively. Here, p_1 divided by p_0 is 1.05, which tells us that the prices increase by (1 + interest rate).

Now, we can extend this calculation to T years. Again, we only show the result of the calculation, based on the following prices: $p_T = (1+r)p_{T-1} = (1+r)^2 p_{T-2} = \cdots = (1+r)^{T-1} p_1 = (1+r)^T p_0$. In other words, we can find out the price after T years by multiplying the price of the resource at the initial point by (1 + interest rate) T times.

Figure 12.5 shows the remaining amount of a resource $\left(= 10 - \sum_{t=0}^{i} x_t \right)$ and its price as the resource is used up over six periods (i.e., 5 years). Again, we assume $\alpha = 0.5$, $r = 0.05$, $X = 10$. Note that the curve representing the prices becomes clearer as the period on the horizontal axis becomes longer.

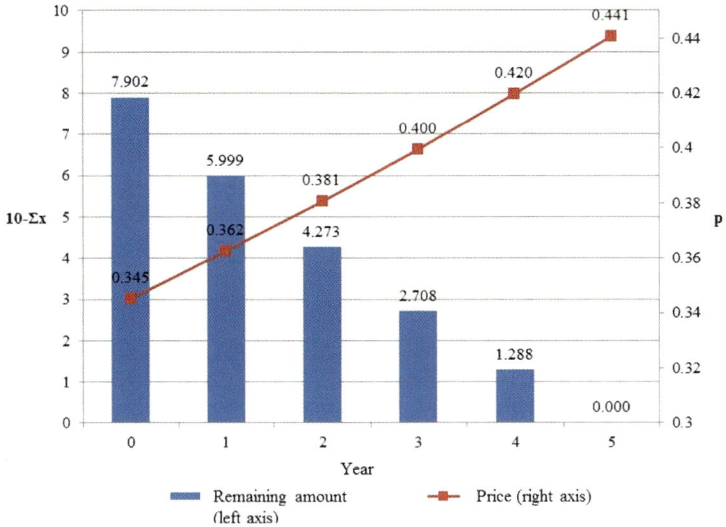

Fig. 12.5 Example of remaining resources and the prices in six periods (Source: Author. Resources at the initial point is 10; interest rate is 5 %)

12.3 The Task of Economics

12.3.1 From Actual Data

Hotelling's Rule was introduced in the previous section as a rational conclusion of economics. When an exhaustible resource is used for a prolonged period, its price will increase at an interest rate as the amount of the resource decreases. The change in the price of an exhaustible resource does not necessarily follow an ascending curve in reality (see Fig. 12.6).

Figure 12.6 shows the confirmed amount of oil reserves and the real prices of imported oil in the United States. The graph is based on data publicly disclosed by the US Energy Information Administration, Department of Energy, on their website. The Consumer Price Index of March 2014 is used to calculate the real price, with the price in 1970 taken as 100. The data of oil reserves are available until 2011, but the price data are available until 2013.

The figure shows that over the 40 years since the beginning of the 1970s, the price of oil has not increased as the supply has decreased.

The price of oil is an important economic index and has an impact on the prices of all kinds of goods. However, this price does not only depend on the available amount of oil reserves. Other factors, such as the existence of alternative energy sources, technology, and the discovery of new oil reserves, also play a role (Krautkraemer 2009).

In resource economics, methods to determine the optimal use of resources have been researched since the 1970s, when the limits of resources and the environment

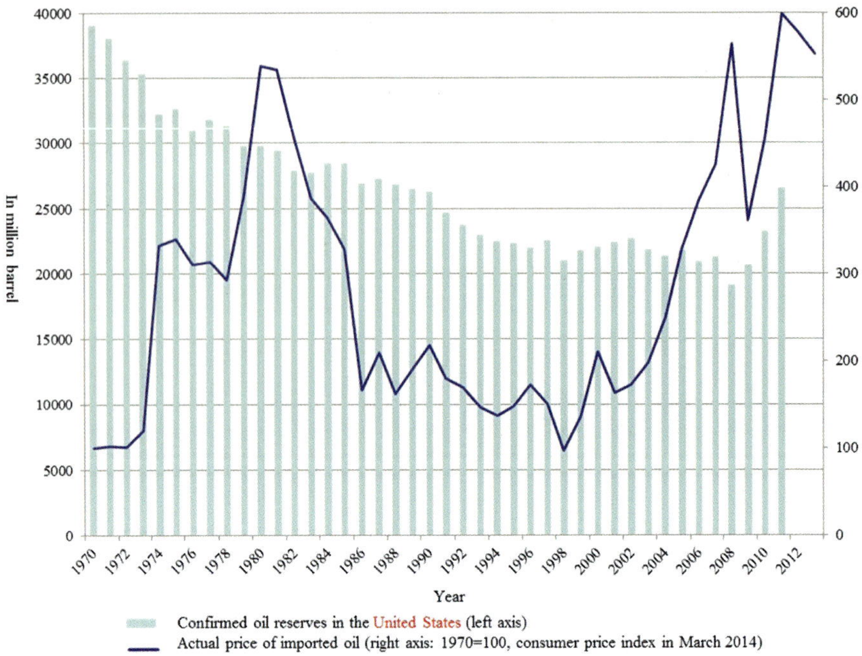

Fig. 12.6 Confirmed oil reserve and prices in the United States (Source: Author: based on publicly disclosed data of the U.S. Energy Information Administration, Department of Energy http://www.eia.gov/)

began attracting attention as a result of the oil crisis at the time. Here, the dynamic optimization model became popular owing to its ease of use.

An effect known as "back-stop technology" was analyzed at much the same time. This required a higher supply cost than oil, but had the same function. Here, the same function as oil can be supplied indefinitely at the same cost if the price becomes equal to the cost (Nordhaus 1973).

However, it is difficult to estimate when such technology appears in an economy. On the other hand, a critical situation that requires alternative technology means economics can be useful when formulating environmental policies.

Japan has begun to emphasize renewable energy, such as solar power and wind power, and is beginning to move away from its dependence on nuclear energy. This shift was triggered by the Great East Japan Earthquake of March 2011 and the subsequent nuclear disaster. However, these alternative sources of energy will take time to replace nuclear power as back-stop technologies.

Are current policies of encouraging renewable energy, such as the feed-in tariff, valid in the long term? These options are expensive to implement. Furthermore, are they going to be better for the environment? Are there any side effects?

The difficulty of predicting answers to these questions means we need to use some trial and error when using these economic models to formulate environmental policies.

12.3.2 Conclusion

In this chapter, we first explained the research fields covered by economics, and noted that mainstream economics does not typically incorporate the natural environment. Next, we examined the optimal use of exhaustible resources as part of resource economics and explained various environment problems from an economics viewpoint using models and examples. In addition, we learnt that economic theories reach conclusions that are easy to understand, but fail to fully explain the real movements of oil stocks and prices.

A sound understanding of real-world situations and environmental problems is required before we can solve these problems using economics. Then, to accurately express the structure of a problem, we require a mathematical model, the results of which should be considered when formulating policies.

While this sounds reasonable, there are many factors to be taken into account and balanced. Most (mainstream) economics based on mathematical principles require that agents make rational decisions. Models that deviate from this assumption tend not be accepted. Moreover, descriptions not based on mathematical principles are considered to be lacking in logic and, consequently, are often ignored.

However, once a precedent is accepted, the range of acceptance of economics widens. A recent example is the field of behavioral economics, which has become most popular (Makabe 2010). However, whether behavioral economics can replace mainstream economics remains to be seen.

In contrast, environmental economics and resource economics still have little influence in economics and in the general community. This chapter briefly introduced resource economics, but readers are encouraged to explore the methods explained here further.

References

Conrad JM (1999) Resource economics. Cambridge University Press, New York

Daly HE, Farley J (2004) Ecological economics: principles and applications. Island Press, Washington, DC

Daly DE (2005) Beyond growth: the economics of sustainable development (trans: Nitta I, Kuramoto S, Omori M). Beacon Press, Boston

Hotelling H (1931) The economics of exhaustible resources. J Polit Econ 39:137–175

Krautkraemer JA (2009) Economics of scarcity: the state of the debate. In: Simpson D, Toman M, Ayres R (eds) Scarcity and growth revisited: natural resources and the environment in the new millennium (trans supervised: Kazuhiro Ueta). Nippon Hyoron Sha Co., Ltd. Chapter 3

Makabe A (2010) An introduction to behavioral economics. Diamond, Inc, Tokyo

Nordhaus WD (1973) The allocation of energy resources. Brook Pap Econ Act 3:529–576

Söderbaum P (2010) Understanding sustainability economics: towards pluralism in economics (trans: Omori M, Koiwai H, Noda K). Shuppanken

Stiglitz JE, Walsh CE (2012) Economics, 4th edn. (trans: Yabushita S et al).Toyo Keizai Inc., p. 22

Chapter 13
Monetary Value of the Environment: Theory and Method

Mitsuyasu Yabe

Abstract Commodities and services traded in markets are priced, and economic activity is conducted when people compare and consider the prices and quality/quantity of products on sale. However, it is difficult to form a market for the environmental commodity because of its non-excludability. That is, it can be used without paying a price. Therefore, it becomes difficult to fulfill supply and demand appropriate for its value, and both undersupply and overuse occur.

As a result, it has become necessary to determine standards for the adequate supply and use of the environmental commodity by means of economic valuation, comparing and considering its value. Here, various methods to quantify the monetary value have been developed. The preference-independent valuation method evaluates environmental value based on scientific data, while the preference-dependent valuation method is based on peoples' preferences. This chapter focuses on the latter method.

Keywords Stated preference method • Contingent valuation method • Classification of environmental value

13.1 Classification of Environmental Values

There are various methods to quantify environmental values, and, as a result, the resulting values also differ. This section discusses the main environmental values that need to be determined. Turner et al. (1993) and Yabe (2014) were referred to for classification and valuation methods of environmental value in this chapter, and Yabe (2007) was referred to for welfare measurement and features of CVM, with addition and modification.

Figure 13.1 summarizes the main environmental values. Each environmental value is evaluated through peoples' preferences, and the total economic value is the aggregate of them all. This total economic value is divided into a use value and a non-use value. A use value is a value acknowledged on the basis of the use of the

---

M. Yabe (✉)
Department of Agricultural and Resource Economics, Kyushu University, Fukuoka, Japan
e-mail: yabe@agr.kyushu-u.ac.jp

T. Shimaoka et al. (eds.), *Basic Studies in Environmental Knowledge, Technology, Evaluation, and Strategy*, DOI 10.1007/978-4-431-55819-4_13

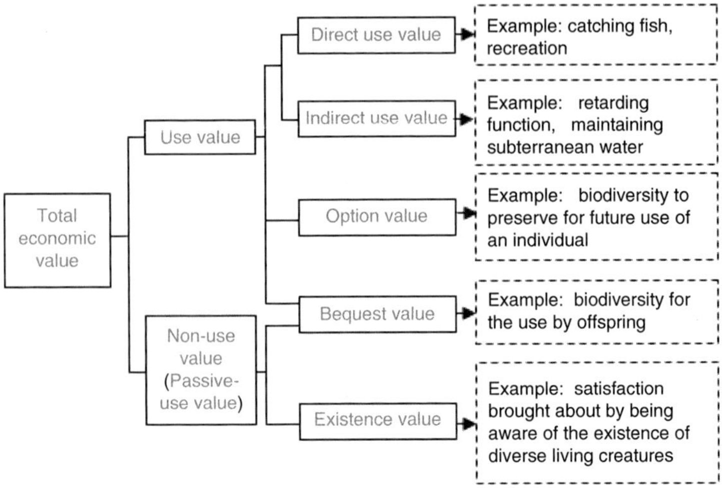

Fig. 13.1 Classification of environmental value

environment. A non-use value (or passive-use value) is a value acknowledged without any such use.

Use values include direct use values and indirect use values. A direct use value occurs when people use the environment directly. For example, in a wetland, the value is gained when people catch fish to make a living or when they use the wetland for recreation purposes. On the other hand, an indirect use value is acquired simply by the existence of the environment, without any initiative for its use being taken by local residents. Examples include the retarding function of a wetland in the event of flooding and the maintenance of subterranean water supplies. Furthermore, there is an option value. An option value is a value acknowledged for potential future use, even if the environment is not currently in use. For example, this may be the amount people agree to pay to have an opportunity (option) to see rare birds on the verge of extinction in the wetland at some point in the future.

Next, non-use value (passive-use value) includes bequest values and existence values. A bequest value is the value associated with leaving the environment for the offspring of a person, even if the person himself has no opportunity to use the environment. For example, a person may make a donation to preserve the biodiversity-rich wetland for his offspring, even though he will never visit the wetland himself. Here, the donated amount is the bequest value of the wetland. Some studies describe bequest values as use values in the sense of the utility produced by people.

An existence value is the value associated with being aware of the existence of the environment. For example, a person or his offspring may gain satisfaction or enjoyment simply from knowing that the diversity of life in the wetland is preserved, even though they will not visit there. Here, the wetland is considered to have existence value.

However, not everyone agrees on these divisions of value. The division of value differs according to the people involved. For example, unknown creatures living in a wetland only have existence value for most people, but they have use value for researchers of genetic resources.

13.2 Valuation Methods of Environmental Value

13.2.1 Concept of Classification

There are various methods for valuing the environment. Six major valuation methods are introduced here. These valuation methods are divided into the preference-independent valuation method, where peoples' preferences are not reflected in the evaluated value, and the preference-dependent valuation method where their preferences are reflected. Then, the preference-dependent valuation method includes the revealed preference method, where the environmental value is estimated based on the relationship between the market value and consumer demand, and the stated preference method, where citizens directly estimate the environmental value via a survey. Table 13.1 indicates the relationship between the environmental values and the valuation methods.

13.2.2 Preference-Independent Valuation Method

Dose-Response Method

In the dose-response method, environmental levels and production are linked. Here, the focus is on the change in market value brought about by changes in environmental standards. For example, the relationship between the deterioration in the environment and a corresponding decrease in the catch of fish may be estimated. Then, this decrease is multiplied by the price of fish, providing the economic loss caused by the deterioration of the environment. Alternatively, the number of fish or birds inhabiting a wetland may be estimated and then multiplied by their price. This provides a preservation value of the wetland.

Replacement Cost Method

Here, the valuation of replacement cost method is based on the cost required to provide environmental services in a similar market commodity. For example, the "retarding function" of a wetland is evaluated according to the construction cost and maintenance cost of a flood control dam with a similar water storage capacity to the wetland in heavy rain. Alternatively, the value may be estimated by the cost of

Table 13.1 Valuation methods and environmental values

Valuation model			Direct use value	Indirect use value	Option value	Bequest value	Existence value
Type		Valuation method					
Preference-independent type		Dose-response method	○	○			
		Replacement cost method	○	○			
Preference-dependent type	Revealed preference method	Travel cost method	○				
		Hedonic pricing method	○	○	○		
	Stated preference method	Contingent valuation method	○	○	○	○	○
		Conjoint analysis	○	○	○	○	○

building, restoring vegetation, and supplying and managing water in a wetland with an equal biodiversity preservation capacity.

Environmental value is measured based on scientific data in the dose-response method and the replacement cost method and not on the peoples' preferences toward the environment. Therefore, these are preference-independent valuation methods. The range in which valuations are available from scientific data has an objective aspect. However, it mainly involves current use. Therefore, the values to be evaluated are limited to direct use value and indirect use value. Furthermore, even though these methods rely on scientific data, the costs vary depending on the quality and quantity of environmental services to be replaced and also include value judgments. In addition, a further problem is that areas that attract many visitors tend not to be valued accordingly, as this demand is not reflected in their environmental value.

13.2.3 Preference-Dependent Valuation Method

Travel Cost Method

Next, we examine the revealed preference methods, which fall within the preference-dependent valuation methods. First is the travel cost method. This method was developed in the United States in the post-WWII period when the demand for recreation increased considerably. The necessity for national and state nature park development increased. This method was used to determine the amount of investment required. In a typical travel cost method, the demand curve is estimated using data on the number of visitors and the costs they incur traveling to the area. The use value is then estimated by the consumer surplus and the travel cost.

Hedonic Pricing Method

The hedonic pricing method is a revealed preference method that quantifies the economic value of environmental factors by estimating the relationship between housing prices/rents and the environmental factors that affect these. For example, when the price of a residential area with many parks is higher than a residential area without any, the value of the parks can be quantified by the difference in the housing prices. Similarly, if the housing price in a noisy area is lower, the damage caused by the noise can be estimated by the decline in housing prices. Since the future utility potential is reflected in the land price, this method can also evaluate option values.

Contingent Valuation Method

With the contingent valuation method (CVM), a virtual situation is assumed, and a survey is conducted on those citizens who are beneficiaries. Citizens are asked to reveal the maximum amount they are willing to pay to prevent environmental deterioration or to promote environmental improvement. In this way, the environment is given a valuation. This method is also known as the stated preference method, since people directly state the environmental value. The merit of the CVM is that the environmental value of the test subject is directly reflected in the evaluated price. In addition, the bequest value and existence value can also be evaluated. Moreover, flexible research designs are available, and valuation is possible irrespective of the function or any preservation policies. On the other hand, the method does have some problems. However, we explain those in a later section.

Conjoint Analysis

Another stated preference method is conjoint analysis. Here, several attributes (e.g., the contribution to preservation activity by birds and plants when evaluating biodiversity) and their levels (e.g., the number of species being preserved and the contribution to this preservation) are combined to present alternative plans. Respondents indicate their order of preference among the alternatives, which are then analyzed. In particular, in the field of environmental valuation, a considerable amount of research has been done on the choice experiment, where one alternative is selected from some alternatives.

Although the choice experiment belongs to the group of stated preference methods, it evaluates the marginal value of change in the environment. This differs from the CVM, which quantifies the value of the overall environment. For example, when the value of biodiversity is evaluated, the marginal value of the increase of a species owing to the improvement in the preservation level of biodiversity is quantified in the choice experiment. In contrast, the CVM determines the value of the overall preserved biodiversity.

As described earlier, many economic valuations of an environment include the replacement cost method, travel cost method, hedonic pricing method, contingent valuation method, and conjoint analysis. Globally, more than half of economic valuations use the CVM. Therefore, we discuss this method in more detail in the following section.

13.3 Contingent Valuation Method

13.3.1 The Willingness-to-Pay and Willingness-to-Accept Compensation

In the CVM, a change in the previous and posterior situations is presupposed. Then, people are asked about their WTP to gain or avoid change or their willingness-to-accept (WTA) compensation to overlook or accept the change. The choice of WTP or WTA depends on what kind of rights is assumed by the researchers toward the environment in which rights are not firmly established.

We begin with a case of environmental improvement. For example, suppose there is a case of water improvement by introducing a purification facility to a factory (posterior situation) on a river polluted by wastewater from the factory (previous situation). When local residents are asked about their WTP for water improvement, they assume they have no right to the posterior situation. This is because they are being asked for their WTP to gain a right in the environment after the improvement, on the assumption that they do not currently have that right. Conversely, if the residents are assumed to have a right in the environment after the improvement, they will be asked for their WTA to overlook the posterior situation.

Next, let us assume a case of environmental deterioration. For example, suppose there is a river with clean water (previous situation) that becomes polluted after a factory is built (posterior situation). If residents are asked for their WTP to avoid water deterioration, they assume they do not have a right to the good situation prior to the deterioration. On the other hand, if they are asked for their WTA to accept the water deterioration, they assume they have a right to the previous situation.

As shown in these examples, the choice of WTP or WTA depends on the researchers' assumptions of societal rights when the rights for the environment are unclear. Moreover, when people are asked about their WTP for environmental preservation, respondents may withhold the answer or may present a protest bid of "not willing to pay" for environmental preservation, even if they recognize the value of the environment. This may happen even when they think they have a right to a good environment. Therefore, it is important to conduct an adequate preliminary survey for the virtual situation and to assume a virtual situation that is less likely to provoke a protest bid.

13.3.2 Diagrammatic Presentation of Welfare Measurement

In this section, the WTP and WTA, the valuation measurements of environmental value used in the stated preference method, are explained in diagram form. For simplification, we use a case with only two commodities, namely, environmental goods and all other commodities, which we call a composite commodity. The price

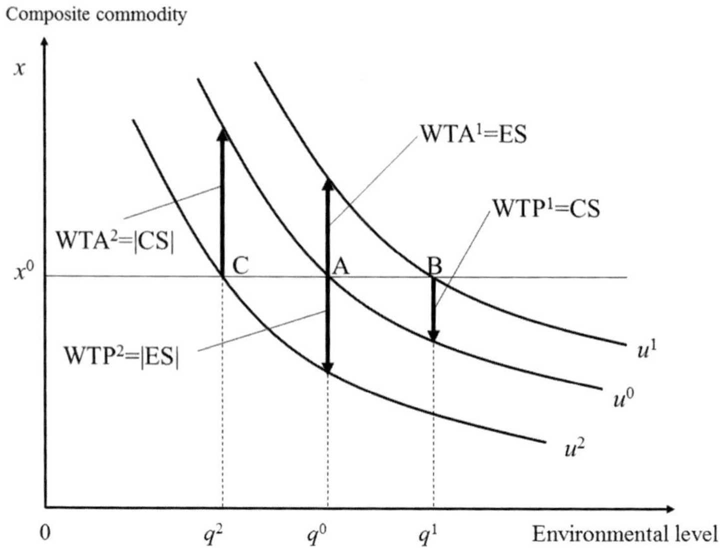

Fig. 13.2 Change of environmental level and compensating surplus and equivalent surplus

of the composite commodity is standardized to 1 (one), and this composite commodity can be considered as money.

Next, the utility function is defined as $u = u(x, q)$. In Fig. 13.2, the x-axis indicates the level of environment q, the y-axis shows the demand for the composite commodity x, and the convex indifference curve u is presupposed to the origin. The indifference curve u indicates a higher utility standard when it is located in the east-northeast.

With regard to the environment, we assume non-excludability because its use is available without paying the price. Thus, the price of the environment is set as zero. This means people can enjoy the benefit of the environment without paying the cost. Moreover, people can only spend their income m on purchasing the composite commodity x. The demand for the composite commodity is derived by dividing the income by its price, $x^0 = m$. Accordingly, the budget constraint line, which indicates the amount of commodity available for purchase within the given income, goes through x^0 and can be drawn as a line horizontal to the x-axis.

When we assume non-rivalry in the environment, meaning that many people can use it simultaneously, the resource constraint does not need to be considered. In addition, the supposition of non-excludability enables people to enjoy maximum use of the environment up to the given environmental level q^0. Moreover, under the given budget and environmental level, the highest utility level u^0 is achieved when the indifference curve passes through the farthest point from the origin in the direction of northeast A (x^0, q^0).

Next, we define the WTP and WTA, given the aforementioned assumptions. First, consider a case in which the environmental level is improved from q^0 to q^1 ($q^0 < q^1$). Consumers use the environment up to q^1 in order to gain higher utility.

The indifference curve shifts to the right, passing through point B and achieving the utility standard u^1 ($u^0 < u^1$).

Consider the maximum amount people are willing to pay to gain the higher environmental level q^1. If they have to pay to gain q^1, the budget constraint line will shift downward, which shifts the feasible indifference curve downward as well, lowering the utility level. In such a case, the utility level can only be lowered as far as the previous utility level of u^0. Therefore, the maximum amount payable becomes WTP^1. As shown here, the adjusted amount of income to bring the utility level back to the previous u^0 under the posterior environmental standard q^1 is called the compensating surplus (CS). Then, the CS is used as the evaluated amount of environmental improvement.

Next, we consider using the WTA to overlook the improvement to the environment. Here, we only need to consider the minimum compensation amount necessary to achieve a posterior utility level u^1 after the environmental improvement under the previous environmental level q^0. This is WTA^1. As shown, the adjusted amount of income to achieve the posterior utility level u^1, based on the previous environmental level q^0, is called the equivalent surplus (ES).

Next, consider a case in which the environmental level is lowered from q^0 to q^2 ($q^2 < q^0$). Then, the indifference curve shifts to the left from point A to point C along the budget constraint line, lowering the utility level to u^2 ($u^2 < u^0$). Accepting a lower environmental level requires sufficient income compensation to achieve at least the previous utility level u^0. That is, a willingness to accept, WTA^2, is needed to gain the previous utility level u^0 under the posterior environmental level q^2. This corresponds to the absolute value of the compensating surplus (CS).

Finally, consider the maximum amount people are willing to pay to avoid environmental deterioration. They will pay money to maintain the previous condition as long as the utility level stays above u^2 under the previous environmental level q^0. The maximum amount is WTP^2, which corresponds to the absolute value of the equivalent surplus (ES). This amount is their willingness to pay to prevent environmental deterioration and to maintain the previous environmental level.

Note that we use the absolute value for the CS and ES, because these take on negative values in the case of environmental deterioration and a lowering of the posterior utility standard (refer to Yabe 2007 for details).

13.3.3 *Major Questionnaire Formats*

Various questionnaire formats have been devised for the CVM, including open-ended, checklist, and dichotomous choice surveys. Many studies have been published on the CVM, including Bateman et al. (2004) and Bateman and Willis (2002). Carson (2007) published comprehensive survey papers. Valuation methods for multifunctionality of agriculture, including biodiversity, are described in Demura et al. (2008).

Details of each questionnaire format are given here. In the open-ended format, respondents provide their own WTP and write it down themselves. Therefore, the questionnaire is simple, but the mental burden on the respondents is considerable. In addition, the response rate tends to be low.

In the checklist format, a table is shown with gradually increasing amounts, starting at zero, and respondents choose the amount corresponding to their WTP or the section that includes the WTP. When the subject is familiar with the respondents, this is relatively easy to answer. However, when the subject is not a daily issue, such as the value of the environment, the response rate tends to be low. In addition, there is a problem of not being able to eliminate the strategic bias when choosing an excessive amount beyond respondents' WTP for environmental preservation.

In the dichotomous choice format, 5–10 questionnaires with different amounts are provided. For example, various monetary amounts are presented such as 5 USD, 10 USD, 20 USD, 30 USD, 50 USD, and 100 USD, and the questionnaires are distributed randomly to respondents. Respondents simply need to answer "will pay" or "will not pay" to the presented amount. In addition, a third choice of "undecided" is often available in accordance with the recommendation of the NOAA report. The environmental value is estimated based on the information of will pay/will not pay and the presented amount. The mental burden on respondents is reasonably light in this case. Therefore, the response rate tends to be higher. However, the method requires a greater number of samples. Furthermore, it has been pointed out that the valued amount is influenced by the method and functional type used for the estimate, although it can deal with strategic bias.

13.4 Conclusion

The CVM has the advantage of being applicable to various situations. However, some researchers find it has bias. For example, the existence of a "protest bid" was mentioned previously. There is also the problem of the embedding effect and the warm glow effect. When the value of the overall environment is not significantly different from the value of a part of it, we have an embedding effect. For example, there may be little difference in the evaluated value when the number of birds observed in a wetland increases from 10,000 to 100,000 or when the range of subjects to be evaluated expands. One reason for this phenomenon is the warm glow effect. That is, people do not indicate their WTP by economically evaluating the value of the environment, but instead indicate their WTP based on their charitable spirit or simple satisfaction of cooperating with environmental preservation. Therefore, it is important to clarify the subject to be evaluated, and the difference between the values and the standards should be explained in an intuitively simple way to safeguard against these kinds of problems.

In addition, the options in a dichotomous choice may include bias. The reason is the starting point effect or the anchor effect, where respondents are influenced by the presented amount when they respond. This is because respondents choose "will pay" beyond their means when there is no other choice to protect the environment, even if they feel the presented amount is too high. Other respondents may believe the presented amount is a general amount. There is also the problem of the yea-saying effect, where respondents too easily say "will pay" to the presented amount. These are some of the reasons for an overvaluation of WTP. However, various ways to deal with the problems have been examined. For example, Yabe (2012) suggests eliminating the presented amount bias by introducing a presented amount effect function. These valuation methods continue to improve as more research is conducted.

References

Bateman I, Willis K (eds) (2002) Valuing the environment preferences: theory and practice of the contingent valuation method in the US, EC, and developing countries. Oxford University Press, Oxford

Bateman I et al (2004) Economic valuation with stated preference techniques: a manual. Edward Elgar Publishing, Cheltenham

Carson RT (2007) The stated preference approach to environmental valuation: applications: benefit-cost analysis and natural resource damage assessment. Ashgate Pub Co, Berlington

Demura K, Yamamoto Y, Yoshida K (authors and editors) (2008) Economic valuation of agricultural environment – multifaceted function, environmental accounting, ecology. Hokkaido University Press, Sapporo

Turner KK, Pearce D, Bateman I (1993) Environmental economics: an elementary introduction. The Johns Hopkins University Press, Baltimore, pp 108–128

Yabe M (2007) Environmental values and contingent valuation method. In: Tokimasa T, Yabuta M, Imaizumi H, Ariyoshi N (eds) Economics of environment and resources. Keiso Shobo Publishing Co., Ltd, Tokyo, pp 162–179

Yabe M (2012) Valuation of preservation value of biodiversity and cultural scenery in Aso Plain. In: Yokogawa H, Takahashi Y (authors and editors) Agricultural formation in bioresearch context and direct payment for environment – approach from agricultural environmental policy theory. Seizansha, Co., Ltd, Sagamihara, pp 173–193

Yabe M (2014) Definition of multifaceted function, policy measures, and economic valuation. In: Yabe M (author and editor) Multifaceted function of grassland farming and animal welfare. Tsukuba Shobo, Tokyo, pp 14–39

Chapter 14
Sustainable Development

Shinji Kaneko

Abstract The concept of sustainable development was first used in the report prepared by the World Commission on Environment and Development (commonly known as the Brundtland Commission), which was established by the United Nations in 1984. It is defined as "development that meets the needs of the present without compromising the ability of future generations to meet their own needs" (World Commission on Environment and Development 1987). Since then, many discussion points and interpretations involving the actualization and elaboration of this abstract concept have been proposed in order to connect it with policies and efforts that aim to achieve such development and progress. One of these efforts involves the development of indicators. The discussion points and indicator development can broadly be reorganized to two origins of argument. The first considers sustainable development to be an extension of discussions regarding how development or progress should be approached (intragenerational equity), while the other regards the concept as an extension of discussions on global sustainability (intergenerational equity). The purpose of this paper is to organize the main discussion points by dividing them into two origins and to gain an understanding of how they are reflected in the development of indicators.

Keywords Sustainable development • Strong sustainability • Weak sustainability • Genuine saving • Ecological footprint

14.1 Introduction

The concept of sustainable development was first used in the report prepared by the World Commission on Environment and Development (commonly known as the Brundtland Commission), which was established by the United Nations in 1984. It is defined as "development that meets the needs of the present without compromising the ability of future generations to meet their own needs" (World Commission on Environment and Development 1987). Since then, many discussion points and

S. Kaneko (✉)
Graduate School of International Development & Cooperation, Hiroshima University, Higashihiroshima, Japan
e-mail: kshinji@hiroshima-u.ac.jp

T. Shimaoka et al. (eds.), *Basic Studies in Environmental Knowledge, Technology, Evaluation, and Strategy*, DOI 10.1007/978-4-431-55819-4_14

interpretations involving the actualization and elaboration of this abstract concept have been proposed in order to connect it with policies and efforts that aim to achieve such development and progress. One of these efforts involves the development of indicators. The discussion points and indicator development can broadly be reorganized to two origins of argument. The first considers sustainable development to be an extension of discussions regarding how development or progress should be approached (intragenerational equity), while the other regards the concept as an extension of discussions on global sustainability (intergenerational equity). The purpose of this paper is to organize the main discussion points by dividing them into two origins and to gain an understanding of how they are reflected in the development of indicators.

14.2 Discussion Points from the Perspective of How Development Should Be Approached

The terms "development" and "progress" are currently understood as being equivocal concepts, since they have been defined or given implications from various perspectives. The discussion on how development should be approached began when industrialized countries accelerated aid to many developing countries that had gained independence following World War II. This created the need for both aid donors and recipients to consider what constitutes development and aid in an internationally comparable way. At that time, the primary objective of development aid was believed to be assistance for developing countries experiencing major economic disparity with developed countries in achieving economic development. Therefore, economic growth was the only development objective, while increasing GDP per capita was the only index used for development from the 1950s to the 1960s. Since "trickle-down economics" (the idea that the benefits of economic growth will eventually reach the entire society, including those in the low-income brackets) was the predominant view with regard to the issues of individual welfare and income distribution and inequality, no political or policy attention was directed toward these issues (UNDP 1990).

Various conditions that were believed to lead to economic development at that time were also viewed as components of development or growth. For instance, urbanization occurs due to the accumulation of capital associated with industrialization after a departure from the traditional agrarian society, leading to the expansion of production and consumption levels. In addition, agricultural productivity improves dramatically, and nationwide infrastructure development, including metropolitan and rural areas, leads to economic progress.

Beginning in the 1970s, however, doubts about the trickle-down theory started to emerge due to persistent poverty and rising income inequality despite the development associated with nationwide economic growth. This led to popularization of the concept of basic human needs (BHNs) as a new measure to provide support that is directly beneficial to those suffering from poverty. BHNs refer to a fundamental element that constitutes the basis for human living, such as food, housing, sanitation, health, and education. It originated from the "New Directions" legislation by the US Agency for International Development (USAID) in 1973. Since then, the objective of the aid provided to developing countries has been to fulfill BHNs for those in the lower-income brackets. This objective was illustrated at the International Labour Organization's (ILO) World Employment Conference in 1976 (The Japan Society for International Development 2014). Moreover, BHNs can be viewed as an essential aspect of quality of life (QOL), which is a concept that has been used for many years in the field of international development in order to improve inhumane living conditions.

In 1990, the United Nations Environment Programme (UNEP) proposed a people-centered development approach and published the "Human Development Report." In addition to the fulfillment of BHNs and the pursuit of QOL, this approach places a higher developmental priority on human development, which encompasses various elements contributing to equal opportunities, gender equality, social participation, and democracy. Economic development was considered an important way to achieve those goals.

The 1992 United Nations Conference on Environment and Development (UNCED), which was held in Rio de Janeiro, adopted the basic principles intended to achieve sustainable development (the Rio Declaration) and its action plan (Agenda 21). While many principles were incorporated in these agreements, for sustainable development they emphasized the need for working collaboratively at every global stage to achieve coexistence between the environment and global development, as well as the importance of solving poverty issues on the premise of "common but differentiated responsibilities." In 1994, John Elkington, the founder of a British consultancy called SustainAbility, proposed the novel idea that corporate activities should be evaluated comprehensively based on the triple bottom line (TBL), which consists of a company's profit and loss account (i.e., a measure of corporate profits), people account (i.e., a measure of social responsibility), and planet account (i.e., a measure of environmental responsibility) (The Economist 2009).

Subsequently, the idea of accomplishing these three aspects of sustainability simultaneously also gained momentum during the process of organizing the concept of sustainable development in global discussions regarding international development. The Johannesburg Declaration on Sustainable Development, which was adopted by the World Summit on Sustainable Development (WSSD) held in Johannesburg in 2002, provides the following description: "We assume a collective responsibility to advance and strengthen the interdependent and mutually reinforcing pillars of sustainable development—economic development, social development, and environmental protection—at the local, national, regional, and global levels."

The 2012 United Nations Conference on Sustainable Development (UNSCD) held in Rio de Janeiro indicated the need to establish a "system" to achieve the coexistence of economic development, social development, and environmental protection. Also discussed was the need to organize the concepts and developing indexes that encompass the following four areas: economy, society, the environment, and the system.

14.3 Discussion Points Regarding Sustainability

The functions and services provided by the global environment are extensive and can broadly be classified into four areas: (1) natural resource base, (2) waste and pollution-absorbing (sink) functions, (3) natural amenity services, and (4) life support functions. In the 1960s, people began to realize the finiteness and limitations of these global environment functions and services. Such discussions can be said to have originated from Buckminster Fuller's *Operating Manual for Spaceship Earth* (1963), which viewed fossil fuels as energy savings and argued the importance of natural renewable energy use (Fuller 1963). In response to this argument, Kenneth Boulding published *The Economics of the Coming Spaceship Earth* (1966). It referred to the conventional economy built on the premise of unlimited resources as the "cowboy economy," and it asserted that it will become important in the future for humans to acknowledge the "spaceman economy" living on the circulating ecosystem and the closed system of the earth (Boulding 1966).

Commissioned by the Club of Rome and published in 1972, *The Limits to Growth* was a research report prepared by Dennis Meadows and his team at MIT. It set a precedent for building a global system model, and it presented long-term predictions of the global system until 2100 using system dynamics modeling (Meadows et al. 1972). Based on the results, Meadows et al. argued that the earth would reach its limits on the following indexes if population and industrial investment continue to grow at the same pace: (1) cultivated land area, (2) land yields, (3) nonrenewable resources, and (4) the capacity of the earth to absorb pollution.

While the finiteness of the global environment started to attract more attention, the United Nations Conference on the Human Environment (commonly known as the Stockholm Conference) was held in June 1972. The motto of the conference was "Only One Earth," and it was the first large international conference to focus on global scale environmental issues, with 113 participating countries (both developed and developing countries). This led to the adoption of the convention concerning the Protection of the World Cultural and Natural Heritage by UNESCO (in November 1972) and the establishment of the UNEP (in December 1972). However, developed countries and developing nations did not agree on whether they should advance conservation of the natural environment or development (poverty reduction). Such disagreements had impacts on the process of establishing the "sustainable development" concept pursued by the Brundtland Commission in 1984.

In general, there are two main views on the principle of sustainable, long-term development, management, and conservation of the global and natural environments. One of the typical examples is described in a concept arguing that "the current generation does not especially owe to its successors a share of this or that particular resource (e.g., petroleum resources). If it owes anything, it owes generalized productive capacity (e.g., productive capacity currently exerted by using petroleum) or, even more generally, access to a certain standard of living or level of consumption" (Solow 1986). These discussions are reflected in the different views on the substitute relationship between natural and man-made capital. A case of the two extremes involves perfect substitutability and perfect non-substitutability. Imperfect substitutability, which excludes non-substitutability (called "weak sustainability"), and imperfect non-substitutability, which excludes substitutability (called "strong sustainability"), can be found in between these two extreme cases.

John M. Hartwick (1977) is known to represent the former case. Exhaustible resources will eventually be depleted, and environmental destruction halts the supply of ecological services. At that point, intergenerational equity no longer remains undisrupted. Speaking of extremes, if technologies are developed that generate economic value from non-exhaustible natural resources, such as sunlight or technologies that enable us to replace the function of maintaining productive capacity and natural purification capacity with increased man-made capital, it would be possible to ensure the level of consumption and standard of living for future generations to a certain extent even if natural resources are depleted or the environment is destroyed. Therefore, Hartwick argued that investing economic profits gained from exhaustible resources (part of natural resources) in productive capital (man-made capital) that generates the same value in the future will solve the problem of having to reduce the levels of consumption for future generations.

On the other hand, the theory of strong sustainability argues that maintaining a certain level of natural capital is an essential condition for sustainability as it is based on the view that natural systems are a mutually dependent, complex mechanism that cannot be understood by humans entirely or placed under complete control. Herman E. Daly proposed three principles: (1) the yield of a renewable resource can be no greater than the rate of its regeneration, (2) the amount of waste and pollutant emissions cannot exceed the rate at which the environment can assimilate and absorb them, and (3) the output of an exhaustible resource can be no greater than the rate at which a renewable resource can be substituted for it (Daly 1990, 1996). This shows that the degree of substitutability with man-made capital or technologies is considered to be limited. Similarly, the four system conditions on sustainability in the Natural Step proposed in 1989 by Karl-Henrik Robèrt, a Swedish medical doctor, are also built upon the theory of strong sustainability, as the conditions state that nature is not subject to systematically increasing (1) concentrations of substances extracted from the earth's crust, (2) concentrations of substances produced by society, and (3) degradation by physical means and (4) that in a society, people are not subject to conditions that systematically undermine their capacity to meet their needs (Robèrt et al. 1997).

14.4 Measurement and Indicators for Sustainable Development

14.4.1 Classification of Indicators

While various discussions on sustainable development were taking place, developing indicators also began in order to evaluate the direction toward which actual sustainable development is moving. At the same time, developing indicators also lead to the embodiment of what sustainable development means in more detail. Various sustainable indicators have been proposed, as many definitions and interpretations exist. This section discusses the main indicators developed and then organizes the concepts and differences.

To begin with, consider the significance of developing indicators on sustainability and evaluating the progress of this effort. First, it helps to express complex, abstract concepts in a more comprehensive manner and increase the general public's interest in sustainable development (playing an advocacy role). Second, it provides the opportunity to conduct a comparison between countries and regions and draw lessons from countries with high scores. Third, it creates the opportunity to set a target for achieving policy objectives, allowing to determine whether a certain policy objective has been met or to check whether the policy has made progress toward or regressed from achieving the objective. Fourth, it can be used as information to determine policy priorities.

Next, since sustainable development is an equivocal concept, it is necessary to consider multidimensional conditions. Thus, the approach taken to build the indicators proposed to date can broadly be classified into two types. One of the types is a nonintegrated approach in which an evaluation is conducted by putting several individual indicators side by side, while the other is based on the development of integrated indicators. Moreover, integrating indicators involves cases that integrate individual indicators unified under a certain unit or cases that integrate nondimensional indexed indicators by giving some type of weighting. Table 14.1 presents a summary of the main indicators that have been proposed to date.

Table 14.1 Classification of sustainability indicators

		Global sustainability	
	Development approaches	Strong sustainability	Weak sustainability
Individual indicators	MDGs	PSR	
Integrated indicators	HDI, QLI, HPI, NNW, ISEW, GPI, Green GDP	ESI, EF	GS

MDGs Millennium Development Goals, *HDI* Human Development Indicators, *HPI* Happy Planet Index, *QLI* Quality of Life Index, *NNW* Net National Welfare, *ISEW* Index of Sustainable Economic Welfare, *GPI* Genuine Progress Indicator, *PSR* Pressure-State Response, *ESI* Environmental Sustainability Index, *EF* Ecological Footprint, *GS* Genuine Saving

14.4.2 Individual Indicators

As typical examples of individual indicators, the table shows the MDGs (originated from the perspective of how development should be approached) and the PSR (originated from the perspective of sustainable global environment). The MDGs were created along with the "United Nations Millennium Declaration" adopted at the UN Millennium Summit held in September 2000 as the objective for an international society in the twenty-first century. They are a set of indicators that must be achieved by the international society by 2015, with target values for each indicator. The MDGs consist of 8 goals, 21 targets, and 60 indicators. Goal 1 is to "eradicate extreme poverty and hunger." Goal 2 is to "achieve universal primary education." Goal 3 is to "promote gender equality and empower women." Goal 4 is to "reduce child mortality." Goal 5 is to "improve maternal health." Goal 6 is to "combat HIV/AIDS, malaria, and other diseases." Goal 7 is to "ensure environmental sustainability." Goal 8 is to "develop a global partnership for development." Progress toward these goals is managed by comparing each goal's degree of achievement for each indicator. Since the MDGs' overall target date is set in 2015, the "Post-2015 Development Agenda" is scheduled to be adopted by the UN General Assembly in September 2015, facilitating discussions on the development of new sets of indicators by 2030 and their target values. It is expected that the number of indicators relating to the Sustainable Development Goals (SDGs) will be added to the MDGs that specialize in poverty and development.

With regard to the PSR, this is not a specific set of indicators but rather a framework to select and analyze a set of indicators. For instance, the OECD Environmental Indicators are an example of a set of indicators that uses the PSR framework. For ten environmental issues, including climate change, the ozone layer, and atmospheric pollution, indicators are selected based on the perspective of "pressure," "state," and "response" to conduct an assessment on sustainability. The European Union (EU) has organized indicators by extending the OECD framework as the DPSIR framework (driving force, pressure, state, impact, and response). In addition, sustainability indicators adopted by each country are organized as a set of indicators by selecting individual indicators that can be measured in the country in the areas of the economy, the environment, society, and the system. The number of indicators, target areas, and the proportion among target areas differ significantly depending on the country's priority or interest levels and data management (Tasaki et al. 2007). Therefore, these indicators are often used as a benchmark to measure the degree of progress in each area of sustainable development in the country. Since many of the PSR indicators deal with environmental sustainability and do not actively advocate a substitution between natural capital and man-made capital, they are classified as strong sustainability indicators.

The advantage of using individual indicators is that it allows for the creation of many detailed indicators. Therefore, it is also feasible to set up indicators that measure the "quality" of sustainable development flexibly, which also creates an incentive to develop new indicators. On the other hand, using individual indicators has its own disadvantages as it does not clearly show how priorities and trade-offs between each indicator are decided, and it is not easy to make a comparison by country. It can be stated that frameworks, such as the PSR, are a means to compensate for the disadvantage stemming from the difficulty in discussing such relationships between each indicator and the trade-offs.

14.4.3 Integrated Indicators

Integrated indicators are either one indicator or a small number of integrated indicators that show the degree of achievement or progress for comprehensive sustainability goals. The greatest advantage is that integrated indicators are easy to understand for policy decision makers and the general public. Moreover, they are often devised to allow for international comparisons by using indicators that can be measured in developing countries. On the other hand, since strong assumptions are made for integration or simplification is performed, disadvantages of indexing become more apparent, as evidenced by the fact that various aspects of sustainable development are disregarded. In addition, understanding an integration method becomes essential in order to comprehend and use integrated indicators correctly. When the integration method is complex, it also becomes difficult to decipher a causal relationship.

The HDI (Human Development Index) is a typical example of integrated indicators derived from how development should be approached. The UNDP releases the results in the Human Development Report (HDR), which has been published annually since 1990. The HDI is calculated by normalizing life expectancy, adult literacy rate, gross enrollment rate, and GDP per capita into an index that ranges from 0 (the minimum value) to 1 (the maximum value) and then adding them after giving a weight of 1/3, 2/9, 1/9, and 1/3, respectively. This is a reflection of the idea that maintaining a balance among three aspects—health, education opportunity, and economic stability—is essential for human development. Moreover, the IHDI (inequality-adjusted HDI) has been used for the calculation since 2010. Comparing the GDP per capita ranking by country and the HDI or IHDI ranking by country makes it possible to examine how appropriately economic capacity is used for human development in each country.

The QLI (Quality of Life Index) and the HPI (Happy Planet Index) are examples of nondimensional integrated indicators derived from how development should be approached. This is an attempt to have subjective judgment reflected in indicators since the HDI does not necessarily evaluate human happiness and satisfaction directly. The QLI aims to utilize the questions used in medical fields to measure quality in terms of individual life and living situations and the indexing method developed by Ferrans and Powers (1985) as development indicators for a country or society. The result released by the Economist Group in England in 2005 is a well-known example. The HPI was proposed by a British think tank called the NEF (the New Economics Foundation) based on similar idea. The HPI is an index calculated by adding together life expectancy and the aggregate result of direct responses to the question about the degree of happiness from about 1,000 people in each country (a subject level of happiness) and then dividing by an environmental indicator (ecological footprint, to be discussed later). The BLI (Better Life Index), released by the OECD in 2011 as a more comprehensive indicator, can also be classified into such indicators.

On the other hand, there are several approaches in which GDP statistics are deconstructed and used as welfare indicators. While GDP is the sum of the "economic value via markets (value added)" by specifying the time (1 year) and space (country), these indicators are a method that aims to incorporate additional elements into the GDP. Specific examples of such indicators include NNW (New National Welfare), ISEW (Index of Sustainable Economic Welfare), GPI (Genuine Progress Indicator), and Green GDP. NNW is calculated by subtracting areas that do not contribute to the improvement of welfare, such as pollution abatement costs, commuting costs, and military expenditures from the GNP while adding leisure, housework, and volunteer activity that are essential to welfare. ISEW and GPI are also known as similar indicators that subtract military expenditures, damage caused by criminal activities, environmental damage, and natural capital degradation from economic affluence. Among these indicators, Green GDP aimed to focus on environmental and natural capital degradation in particular and establish the System of Environmental-Economic Accounting (SEEA) by conducting economic evaluations. It was the indicator that the UN emphasized to implement after the Rio Summit in 1992. However, the effort did not go beyond the investigation and research phase due to many obstacles with regard to the assessment of the environment and natural capital degradation. As a result, it never materialized into a system that warrants continuous efforts.

The ecological footprint is a typical example of integrated indicators for strong sustainability. It became well known due to results from a series of research conducted by William Rees and Mathis Wackernagel in the early 1990s. It aims to conduct an evaluation by expressing the earth's environment-carrying capacity in units of land area by different types and calculating how much land area would be necessary for our life and existence. It specifically measures the ecological capital essential to producing all resources to be consumed or absorbing the waste generated, during a certain period (normally for 1 year) for each country or the entire world, and it targets forests, wetlands, and water bodies that are necessary to absorb and purify carbon dioxide and environmental pollutants emitted at farmlands, grasslands, and fisheries for food production. The total area of these lands is defined as the ecological footprint. It is then used to evaluate whether overshoot is occurring by showing how much larger the total land area is compared to the actual land area (environmental carrying capacity).

On the other hand, there are weighted composite indicators in which various environmental pollution indicators are gathered, and experts determine the importance of each indicator to global environment sustainability. The Environmental Sustainability Index (ESI) is a well-known indicator that was devised by environmental experts from Yale University and Columbia University in the United States.

In particular, the ESI is a composite index integrated after many indicators are classified into broad categories, such as environmental systems, reductions of environmental stresses, reductions of human vulnerability, societal and institutional capacity, and responsibility toward global common goods. The ESI indicators have been changed to categories such as reductions of environmental stresses on human health, promotion of ecosystem vitality, and the degree of achievement in sound natural resource management in recent years, making progress and evolving toward the Environmental Performance Index with an integration method that incorporates principal component analysis.

There are only a few integrated indicators based on the concept of weak sustainability. Genuine Saving (GS) is a typical example of such indicators. With the idea that the total of man-made capital stock and natural capital stock will not decrease as its condition for sustainability, the GS aims to evaluate its feasibility with the rate of savings. This is based on the premise of substitutability between man-made capital and natural capital. Pearce and Atkinson (1993) conducted the first research that performed actual measurements based on this concept. Subsequently, the World Bank has been conducting measurements as adjusted net savings. Adjusted net savings are calculated by subtracting the depreciation of man-made capital, the depreciation of various natural capital (the rent of developed natural resources), and damage caused by environmental pollution (carbon dioxide) from gross national savings, adding education expenditures (an increase in man-made capital) to the net national savings and then dividing it by gross national income (GNI).

References

Boulding KE (1966) The economics of the coming spaceship earth. In: Jarrett H (ed) Environmental quality in a growing economy: essays from the sixth RFF forum. John Hopkins University Press, Baltimore, pp 3–14

Daly H (1990) Commentary: toward some operational principles of sustainable development. Ecol Econ 2:1–6

Daly H (1996) Beyond growth: the economics of sustainable development. Beacon Press, Boston

Ferrans C, Powers M (1985) Quality of life index: development and psychometric properties. Adv Nurs Sci 8:15–24

Fuller B (1963) Operating manual for spaceship earth. E.P. Dutton & Co., New York

Hartwick JM (1977) Intergenerational equity and the investment of rents from exhaustible resources. Am Econ Rev 67:972–974

Meadows DH, Meadows DL, Randers J, Behrens WW III (1972) The limits to growth: a report for the club of Rome's project on the predicament of mankind. Universe Books, New York

Pearce D, Atkinson G (1993) Capital theory and the measurement of sustainable development: an indicator of weak sustainability. Ecol Econ 8:103–108

Robert K-H, Daly H, Hawken P, Holmberg J (1997) A compass for sustainable development. Int J Sustain Dev World Ecol 4(2):79–92

Solow RM (1986) On the intergenerational allocation of natural resources. Scand J Econ 88 (1):141–149

Tasaki T, Kameyama Y, Hashimoto S, Moriguchi Y, Harasawa H (2007) The status of formulating sustainable development indexes and usability in the long-term vision scenario research. Proceedings of the 35th annual meeting of environmental systems research, pp 269–276

The Economist (2009) Triple bottom line. http://www.economist.com/node/14301663. Accessed on 31 Jan 2015

The Japan Society for International Development (ed) (2014) Lexicon of international cooperation, 4th edn. The International Development Journal Co., Ltd.

UNDP (1990) Human development report 1990: concept and measurement of human development. United Nations, New York, p 104

World Bank (1992) World development report 1992: development and the environment. Oxford University Press, New York

World Commission on Environment and Development (1987) Our common future. Oxford University Press, New York

Part IV
Environmental Strategy

Chapter 15
Basics in Environmental Economics and Environmental Policies

Toshiyuki Fujita

Abstract This chapter explains the basic concepts in the economic analysis of environmental problems. First, Sect. 15.1 is an overview of the relationship between the environment and society, followed by Sect. 15.2, which uses graphs to explain the cause of environmental problems from an economics viewpoint. The environment tends to be exploited because it is not adequately priced owing to the lack of trading in the market. In order to prevent this exploitation, it is necessary to build structures or systems to inform people of the value of the environment (i.e., environmental policies). Section 15.3 covers the optimum pollution standard as a desirable aim of environmental policies, while Sect. 15.4 introduces practical environmental policy procedures.

Keywords Market equilibrium • Market failure • Marginal abatement cost • Direct regulation • Pigovian tax • Emission trading

15.1 Natural Environment and Human Beings

Many studies have issued warnings about worsening environmental problems. Environmental problems vary in scope, from local environmental destruction (e.g., air pollution caused by automobile emissions and waste problems) to global environmental destruction (e.g., ozone layer destruction and climate change). What they all have in common is that they are caused by human economic activities. Solutions to these problems are imperative, because environmental destruction or a worsening environment damages human society.

The definition of environment is the "outside world that surrounds human beings and reciprocally interacts with them" (Ueta 1996), building a foundation of living for mankind. Its function is generally considered to be the source of natural resources, providing amenity (comfort), and absorbing pollutants, providing various benefits for humans. Environmental destruction occurs when a function is harmed. Therefore, environmental destruction appears in the form of resource exhaustion, destruction of nature, air and water pollution, and so on.

T. Fujita (✉)
Faculty of Economics, Kyushu University, Fukuoka, Japan
e-mail: tfujita@econ.kyushu-u.ac.jp

T. Shimaoka et al. (eds.), *Basic Studies in Environmental Knowledge, Technology, Evaluation, and Strategy*, DOI 10.1007/978-4-431-55819-4_15

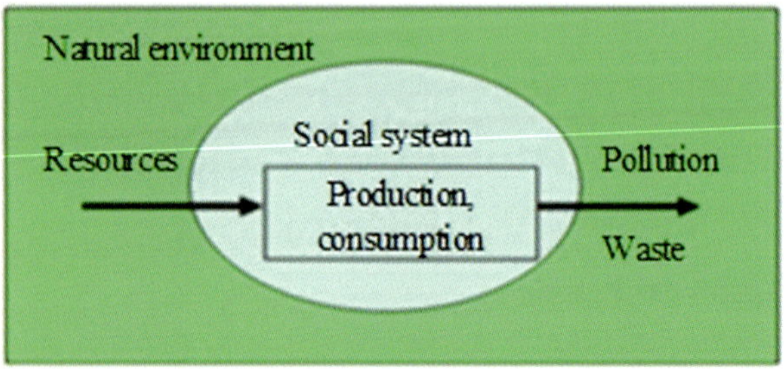

Fig. 15.1 Relationship between nature and society

The human social system has always had an impact on the system of nature. We tend to think that environmental problems began after the Industrial Revolution. However, they had been taking place earlier, but on a local scale. Examples include salt damage in Babylonia caused by irrigation in order to obtain agricultural water or forest destruction due to expansion of agricultural land, which took place with the growing population in Europe (Hosoda and Yokoyama 2007).

Figure 15.1 shows the schematic relationship between nature and society. Society collects resources from nature and receives amenities. Resources are consumed after being converted to products as factors of production and return to nature in the form of pollutants or waste (of course, they sometimes go back into the production process after being recycled). If the range of resource collection is not excessive, nature can assimilate or absorb waste. Recyclable resources, such as forest and biological resources, can be reproduced, maintaining the balance between nature and society.

However, in modern times, with the expansion of the scale of human activities, increased levels of environmental destruction were evident on a global scale at the end of the twentieth century. Global environmental problems became obvious, which indicates that the earth is finite. In 1972, using a simulation based on the "Limits to Growth" model, the Club of Rome predicted that human society would receive devastating damage as a result of resource exhaustion, a food crisis, and pollution if growth continued at the same rate. They stressed that we should stop growth and should strive for a sustainable and stable world. However, the "Limits to Growth" model has received much criticism, and we cannot use it as our future image. Still, many fear that human activities may exceed the environmental capacity of the earth in the near future, considering the trends following the Industrial Revolution. That is, a crucial issue is whether humans can control their tendency to destroy nature and, thus, build a rich society where the environment is preserved.

15.2 Market and Environment

15.2.1 Agenda of Environmental Economics

Here, the function of the market is explained, as well as the reason why exploitation of the environment takes place if the market is not controlled. First, we explain economics, as well as the purpose and issues of environmental economics.

Economics seeks a system that efficiently (i.e., without waste) uses scarce resources. Here, resources include assets, labor (human resources), money, and time, as well as resources such as coal, oil, and minerals. Scarcity refers to the situation in which the desired amount (demand) exceeds the available amount (supply). We have an unlimited amount of air. Therefore, ignoring the quality of the air for now, there is no scarcity of air, and it is not traded on the market. If somebody tried to sell air, nobody would buy it.

In environmental economics, the economic methodology is applied to the development and management of the environment. The environment used to be an abundant asset that we could use freely, much like air, and we did not have to worry about restrictions. Examples of environment use include generating waste and industrial plants emitting pollution in the production process. Today, the environment has become a scarce resource that we need to preserve. Moreover, using the environment comes at a cost. However, there is no market in which to trade the environment, which means there is no adequate pricing. Simply put, we can use the environment for free. For instance, there is no cost associated with emitting pollution and, inevitably, the environment is exploited. This phenomenon is referred to as a market failure by external cost, which will be explained in detail in Sect. 15.2.3.

Government intervention or environmental policies are necessary to correct this market dysfunction and to provide an incentive for environmental preservation. The purpose of environmental economics is to evaluate environmental policies. The agenda of environmental economics is theoretical and needs empirical analyses of policies to realize efficient use of the environment and to provide the information necessary to support these policies.

15.2.2 Market Equilibrium

Here, we explain demand, supply, and market equilibrium, which are basic concepts of economics.

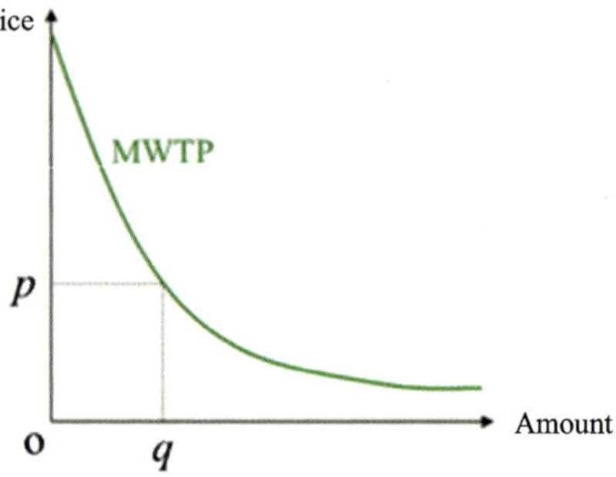

Fig. 15.2 Marginal willingness to pay

Demand

When an individual (consumer) purchases an asset, he or she must see some value in the asset. The value is measured by the consumer's willingness to pay (WTP). When Mr. A orders a cup of coffee for 300 yen at a coffee shop, we know that his WTP for a cup of coffee exceeds 300 yen. Generally, the greater the asset consumption becomes, the lower the value becomes with additional consumption. Value produced by additional consumption is called the marginal WTP (MWTP). Thus, the WTP for a second cup of coffee is lower than that for the first cup, and the WTP for a third cup is even lower than that for the second cup. Assume that Mr. A has a second cup of coffee and that this was his last cup. The MWTP for the second cup is also over 300 yen, but we know the MWTP for the third cup was less than 300 yen. The WTP depends on the individual and on the situation in which the individual is placed.

Generally, the graph of the quantity of assets consumed and the MWTP resembles the one shown in Fig. 15.2, which is actually a demand curve.

In a perfectly competitive market, consumers are price takers (i.e., they are unable to control prices). When p, the price of an asset, is given, they purchase the asset only when their MWTP exceeds p. Therefore, they purchase only q. In other words, the MWTP graph indicates the relationship between price and purchase amount, which is why we call it a demand curve.

The area of $a+b$ in Fig. 15.3 indicates the total WTP for consumption q of an individual (i.e., the value that can be obtained). Furthermore, the area a, calculated by subtracting the paid amount b from the value, is called the consumer surplus. These amounts are all measured as monetary values.

Supply

Companies (producers) produce assets using factors of production, such as labor, capital, and land. It costs money to produce assets. The cost is the amount of money required to purchase factors of production, such as wages for employees, leasing

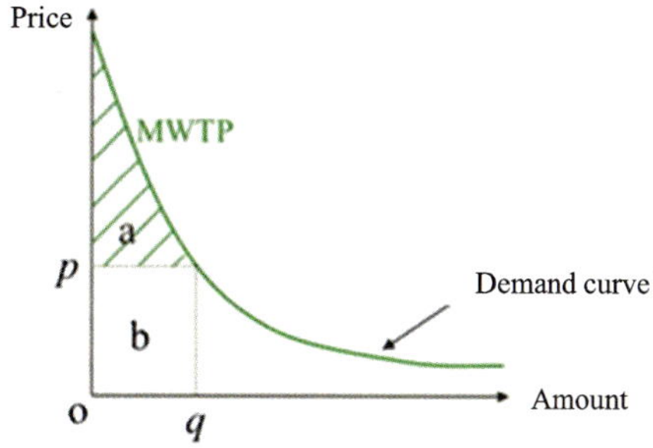

Fig. 15.3 Demand curve

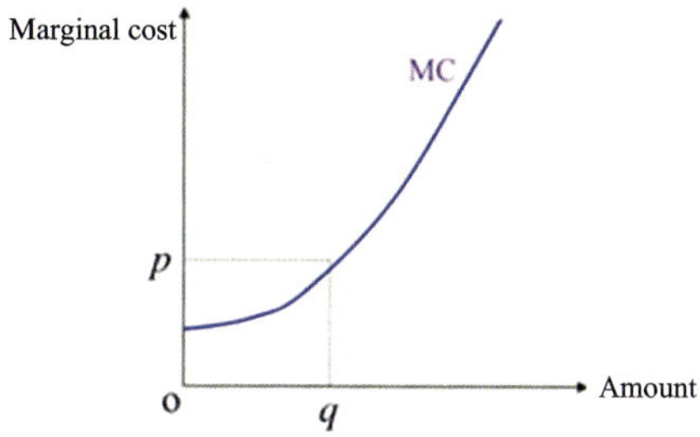

Fig. 15.4 Marginal cost

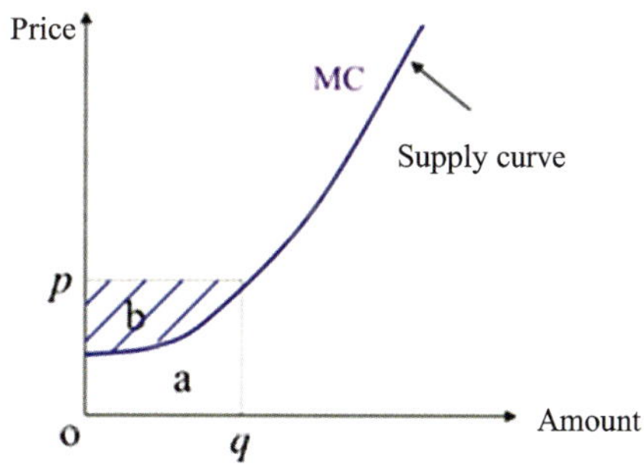

Fig. 15.5 Supply curve

fees for capital equipment, and real estate. As the production of assets increases, the additional cost (marginal cost) to increase production increases at the same time.

Figure 15.4 shows the relationship between output (production amount) and marginal cost. The vertical axis is the price of the asset. Here, Fig. 15.4 is a supply curve. As in the case of demand, when a company is a price taker, it maintains production only in the range where marginal cost is lower than the price. The cost of producing one additional unit is greater when the marginal cost exceeds the price. Therefore, when price is p and production is q, the graph of marginal cost shows a supply curve.

In Fig. 15.5, area a is the total cost, while the area of $a+b$ represents sales. Therefore, the area b is the profit of a company. Profit is also called producer surplus.

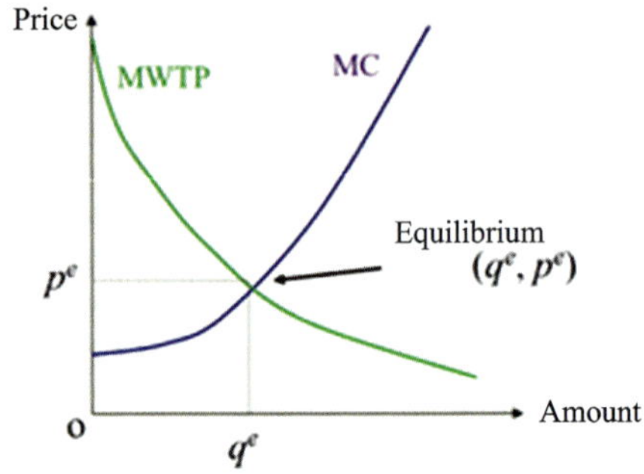

Fig. 15.6 Market equilibrium

Equilibrium

Suppose we have a market in which there are multiple producers and consumers. The demand for the society is the sum of the demand by each consumer. Therefore, the demand curve for society is the horizontal summation of an individual demand curve. The same is true for the supply curve. The amount and price of an asset at the intersection point of the demand curve and supply curve for the society, as shown in Fig. 15.6, is called a market equilibrium. In equilibrium, demand and supply are equal, and there is no change in prices. When demand exceeds supply, there are consumers who wish to purchase even if the prices are higher than they are now, and prices increase. In contrast, when supply exceeds demand, the surplus of unsold stock lowers prices. Through these market adjustment processes, equilibrium is maintained.

Efficiency in the Market

In a perfectly competitive market, the equilibrium point represents the efficient production level and pricing of assets. In economics terms, increasing one person's welfare (surplus) means decreasing somebody else's welfare. In equilibrium, the social surplus is maximized. Here, social surplus is the sum of consumer surplus and producer surplus and is calculated by subtracting total cost from the total WTP.

In Fig. 15.7, the combination of price and amount (p^e, q^e) is in equilibrium, and the area of $a+b$ is the total social surplus. Let's denote $a+b$ as c. As shown in Fig. 15.8a, when the traded amount is q^1, which is smaller than q^e, the sum of the consumer surplus and producer surplus is the total WTP less the total cost (i.e., $c-d$). As is shown in Fig. 15.8b, when the traded amount is q^2, which is greater than q^e, the marginal cost per production unit in the range exceeding q^e is greater than the marginal value, and the social surplus is again $c-d$. In other words, when we underestimate the traded amount or there is excessive trading, the social surplus is less than it is in equilibrium, resulting in a social loss.

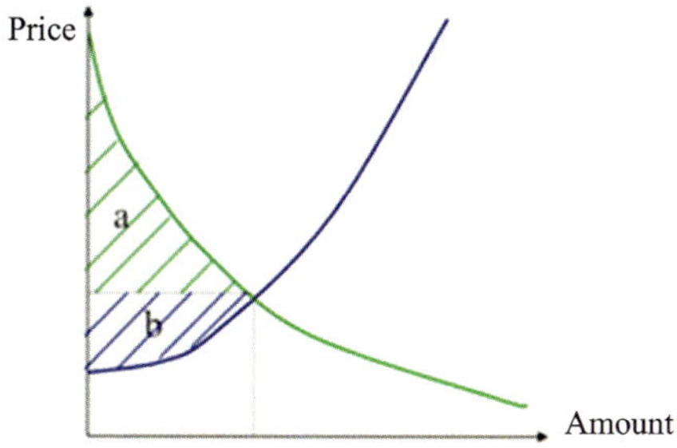

Fig. 15.7 Social surplus

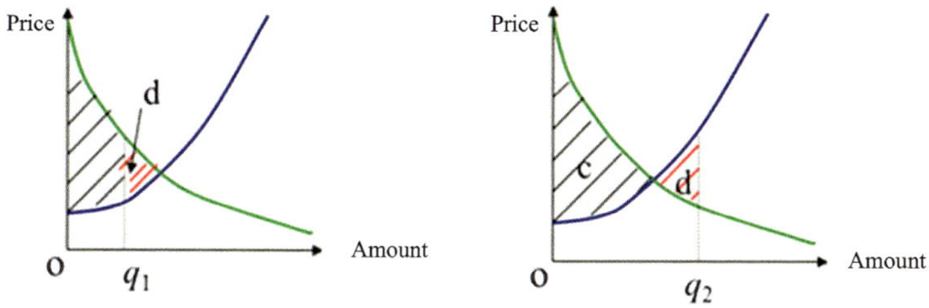

Fig. 15.8 **a** When traded amount is underestimated and **b** when traded amount is excessive

From the above, the optimum condition is realized by letting the market decide on amounts to produce and prices in a perfectly competitive market, with no government intervention.

15.2.3 *Market Failure*

When the production of assets causes environmental deterioration, the market efficiency described in Sect. 15.2.2 is compromised, unless companies consider the social damage brought about by environmental degradation. When we denote this damage as a monetary value, we call it an external cost. Although assessing external costs is extremely difficult, we assume here that it is possible.

Suppose a factory discharges wastewater into a river, which pollutes the water and affects residents downstream. Figure 15.9 shows a marginal social cost graph, which curves up and to the right. This is the marginal cost to the factory (i.e., marginal private cost) plus the external cost (i.e., marginal external cost) of water pollution caused by additional production at the factory. That is, the marginal social cost is the cost to society of producing one additional unit of an asset at the factory.

Equilibrium occurs when the market decides the quantity produced and the price of an asset produced at this factory (i.e., q^e, p^e). A socially efficient condition occurs when the marginal social cost is equal to the MWTP (i.e., q^*, p^*) (see Fig. 15.10).

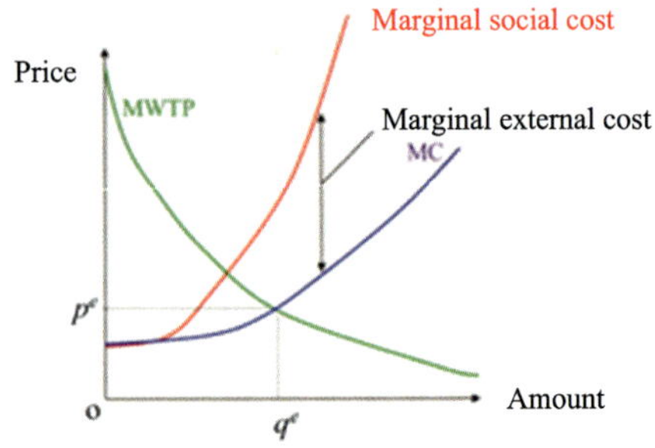

Fig. 15.9 Marginal social cost

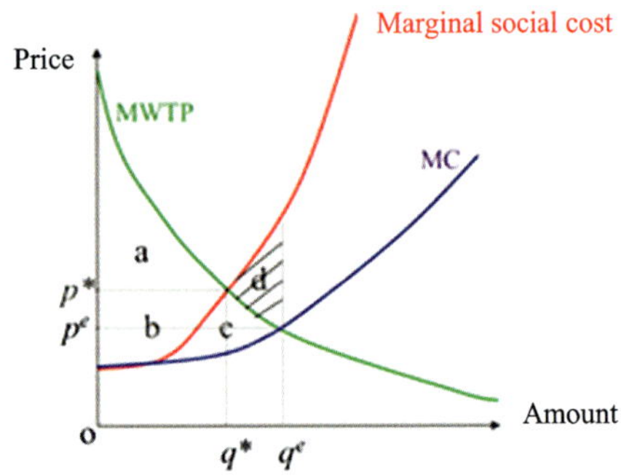

Fig. 15.10 Market failure

This creates a divergence between the market equilibrium and efficient condition. Next, we assess the loss brought about by this divergence. The social surplus in equilibrium, without external costs, is the area of $a+b+c$, as discussed earlier. However, the production of q^e generates an external cost equivalent to the area of $c+d$, creating a social surplus of $a+b-d$. On the other hand, the social surplus at (q^*, p^*) is $a+b$. Therefore, the social loss, when companies fail to consider external costs, is the shaded area of Fig. 15.10 (d). In addition, Fig. 15.10 shows that, in equilibrium, asset production with an excessive external cost means the price is underestimated.

External costs that cause social loss is an example of a phenomenon called market failure. In the context of economics, market failure is the source of environmental problems. In other words, excessive pollution is emitted when the cost of environmental use is not properly recognized. Policies are needed to correct market failures and to create an efficient condition (i.e., a system that can incorporate external costs).

15.3 Cost–Benefit Analysis of Environmental Policies

Section 15.2 indicated that a market cannot achieve an optimum condition with the existence of the external cost, namely, environmental destruction. Here, government intervention is required to correct this market failure. This intervention occurs via environmental policies. Environmental policies consider the amount of pollution emission, such as CO_2 and sulfur oxide. Environmental problems cover a wide range of issues. However, waste and biodiversity cannot be attributed to pollution, but we use the term to refer to all human activities that contribute to environmental deterioration. Emitting pollution has a benefit, but also comes at a cost. Thus, there is a trade-off accompanying increasing emissions, just as with other economic activities.

The expression "benefit of emission" is easily misunderstood. This does not mean that some people derive happiness from pollution. Instead it refers to the social surplus they gain from producing and consuming the assets that caused pollution. Pollution is generated as a result of productive activity, and the assets that are produced have value. The cost of emission is an external cost of pollution (i.e., damage). Assessing the external cost in monetary terms is not easy and is part of a field of study called environment assessment. Environmental assessment tries to assess the financial compensation needed by humans for them to accept the environmental deterioration.

Before continuing, let's clarify the benefits and costs of emission. The social aim is to achieve the volume of emission that provides optimum net benefits (i.e., benefit–cost). Thus, marginal benefit of emission = marginal cost of emission in equilibrium. Usually, the marginal benefit is a decreasing function of emission, while marginal cost is an increasing function. Therefore, the volume of emission that satisfies the condition does exist, in theory. In simpler terms, pollution emissions have positive attributes. The marginal benefit indicates an increase in the good side, while marginal cost indicates an increase in the bad side. Therefore, it is better to increase emissions when the marginal benefit is greater than the marginal cost and vice versa. When both are equal, it is the most efficient.

Next, we consider the above from the viewpoint of emission abatement. Emission abatement means increasing the amount by which we reduce emissions. Thus, reviewing and reducing emissions are similar. In practical terms, the benefit of abatement is an external cost that can be avoided with emission abatement. The cost of abatement is the benefit lost owing to the abatement. Strictly speaking, this is the benefit that would have been gained if the resources assigned to reduce pollution had been used elsewhere at an optimum level. The marginal benefit of emission is the marginal cost of abatement. The increased amount of benefit when emission volume $e+\Delta e$ which is slightly increased from a standard e is almost the same as the reduced amount of benefit $e-\Delta e$ which is slightly less than e (i.e., the cost). Therefore, the condition for optimization can also be shown as marginal abatement cost (MAC) = marginal external cost (MEC).

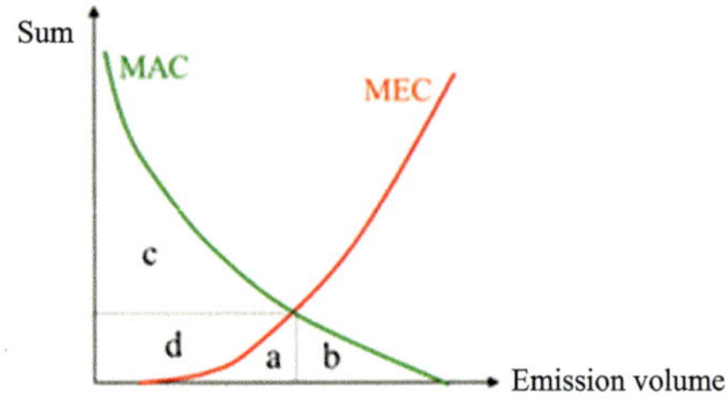

Fig. 15.11 Optimum estimation

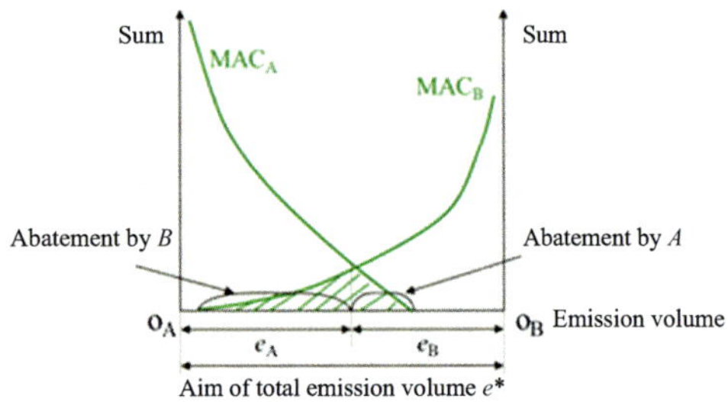

Fig. 15.12 Equi-marginal principle

Figure 15.11 shows the emission volume and marginal benefit and the marginal external cost of emission. When there is no regulation, emission volume is e_0, where marginal benefit (= marginal abatement cost) is zero. Then, the optimum emission volume is e^*, which satisfies the above condition. Alternatively, the emission volume e^* can be interpreted as the abatement volume e-e_0, when the emission volume e_0 is the standard. The sum of total damage (area of a) and the total abatement cost (area of b) is minimized at e^*. Then, benefit–damage is the area of $c+d$, which is maximized.

In the preceding discussion, emission sources in society were all assumed to be combined and regarded as one single source. However, various pollution sources usually exist in diverse ranges. In particular, with the global warming problem, companies and individuals can be considered to be sources of pollution. When multiple pollution sources exist, the marginal cost of all emission sources must be equal to achieve an abatement target at a minimum cost. This is explained in Fig. 15.12.

When two emission sources A and B exist, MAC_A and MAC_B indicate the marginal abatement cost of the emission sources of A and B, respectively. The starting point in graph B is O_B. Note that MAC_B curves down and to the left as the emission volume of B increases to the left from O_B. When the government regulates the sum of emissions A and B as e^*, emissions A and B are set as e_A、e_B ($e^*=e_A+e_B$) so that the marginal costs of both emission sources become equal. The total cost is minimized (the shaded area of Fig. 15.12), proving efficiency. This is known as the equi-marginal principle and is always effective regardless of the number of emission sources.

Intuitively, according to this principle, when two emission sources exist with different marginal abatement costs, the total abatement cost can be reduced without altering total emissions. This occurs by shifting the abatement amount from a higher marginal abatement cost to a lower one by modifying the distribution of the regulation. This type of adjustment is not possible when the marginal cost of all the emission sources is equalized.

15.4 Method of Environmental Policies

15.4.1 Direct Regulation

Specific contents of environmental policies are explained in this section. Section 15.4.1 explains direct regulation, while Sects. 15.4.2 and 15.4.3 discuss the Pigovian tax and emission trading, both methods of economics.

When the government requires companies to legally comply with environmental standards, this is direct regulation and is sometimes called command-/control-type regulation. It is a centralized regulation and a simple and clear policy method. For pollution problems in developed countries, direct regulation was the main method used in environmental policies. Emission standards, technical standards, and discharge rate standards are all included in environmental standards. The discharge rate imposes an upper limit on pollutant emission per production unit.

Direct regulation works to (1) derive a socially optimum total emission volume and (2) distribute total emissions to each emission source (company). In (1), the emission volume is chosen to satisfy the following formula: marginal benefit (marginal abatement cost) of pollution in the whole society = marginal damage of pollution, as explained in Sect. 15.3. In (2), emissions must be distributed to equalize the marginal abatement cost of each company, also explained in Sect. 15.3. In order for a government to achieve these goals, the marginal damage of pollution and marginal benefit of each company must be ascertained. This necessarily means working with a vast amount of information. In addition, when there are many polluters, satisfying the equi-marginal principle is an unrealistic expectation, making efficient regulation difficult. In technology designating regulation, this can potentially harm any incentive for companies to develop reduction technology.

15.4.2 Pigovian Tax

A Pigovian tax is a policy that charges a marginal abatement cost (marginal external cost) in the optimum emission standard (e^* explained in Sect. 15.3) as a tax per emission unit. This term is derived from the British economist who originally devised it, namely, Arthur Pigou. In production activity, the actor using the environment must pay the tax. Thus, imposing such a tax is equivalent to pricing

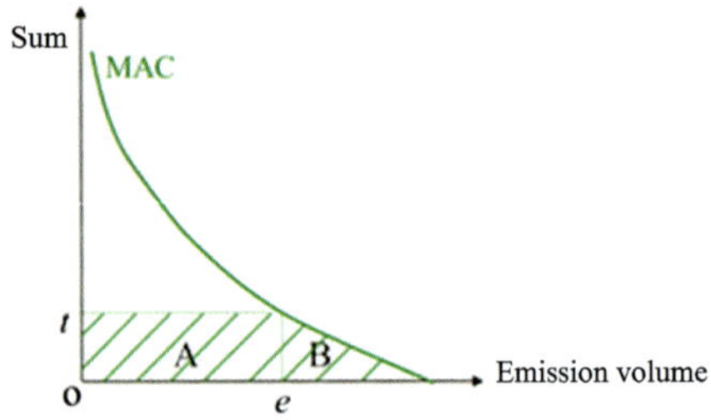

Fig. 15.13 Pigovian tax and behavior of companies

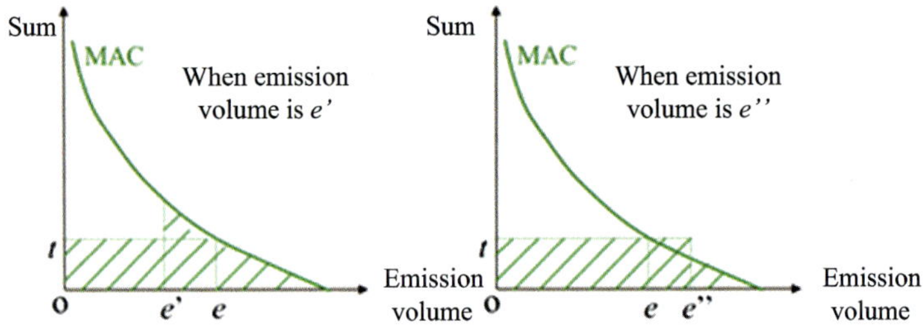

Fig. 15.14 Cost of companies under different emission volumes

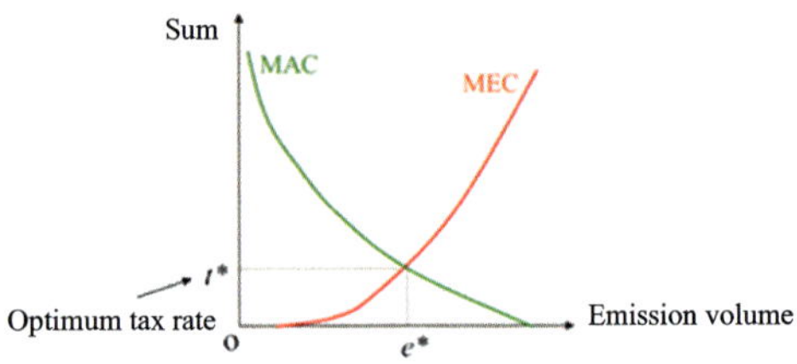

Fig. 15.15 Optimum Pigovian tax rate

the environment. A Pigovian tax is a policy method designed to give companies an incentive to reduce emission.

Imagine a company facing a Pigovian tax. When the tax rate t is given, the company would rationally choose emission volume e to make the marginal abatement cost t, which would minimize the sum of the tax payment and the abatement cost (total cost). In Fig. 15.13, A is the tax payment and B is the abatement cost. Figure 15.14 shows that total cost is higher when a company chooses an emission volume other than e.

Therefore, the Pigovian tax rate is equal to the marginal abatement cost t^* corresponding to optimum emission standard e^* (see Fig. 15.15). Note that the company faced with tax rate t^* does not choose e^* because of his own rational behavior.

When the same rate of Pigovian tax is imposed on multiple companies, the marginal abatement cost of each company becomes the same as the tax rate. Therefore, a Pigovian tax equalizes marginal expenses. In this respect, a Pigovian tax is better than direct regulation and is appropriate as regulation when numerous unspecified emission sources exist, such is the case with global warming. However, similar to the case with direct regulation, a vast amount of information is required to decide on an optimum tax rate. The government does not need to know the abatement cost of each company. However, it is necessary to grasp the abatement cost in the area or the industry that it regulates. Furthermore, information on pollution damage is necessary. Therefore, the Pigovian tax has never been implemented (Ueta et al. 1997). Theoretically, a Baumol–Oates tax is more practical. Here, a reduction target is tentatively set, and the goal is achieved at minimum cost by modifying the tax rate in a trial-and-error fashion.

Policies in Japan, which are similar to a Pigovian tax, include charging for waste disposal bags and industrial waste tax. The charged waste disposal bags is a policy requiring residents to place domestic garbage in designated charged bags and is used in many municipalities. Those who generate waste pay a set price per bag according to the amount of waste. Moreover, an industrial waste tax of 1,000 yen per ton of waste has been introduced in many prefectures. However, both tax rates are too low, so these should be considered as methods to procure a finance rather than methods to reduce the amount of waste.

There is a policy where grant money per unit of waste reduction is given to a company instead of charging tax. This is called a Pigovian subsidy policy. The eco-point system to promote the purchase of replacement appliances with better energy efficiency can be considered as a type of Pigovian subsidy policy for households.

The deposit/refund system is a combined policy of tax and subsidy. One example of this method is the empty can regulation in order to regulate the deterioration of scenic areas due to the unlawful dumping of empty cans. A certain surcharge (deposit) is placed on the price of canned drinks. Consumers who return empty cans to designated sites receive the deposit. Dumping empty cans causes an external cost, namely, that of scenic deterioration. The empty can regulation provides two incentives to consumers to reduce empty can dumping: (1) to reduce consumption of empty cans by charging a deposit and (2) to reduce empty can dumping by returning the deposit.

15.4.3 Emission Trading

Emission trading is a system in which governments decide on a desirable total volume of pollutant emission. This emission allowance (or emission certificate) is distributed to each company. Following this, companies may negotiate the emission allowance. When the government allocates an emission allowance to each company, this policy is the same as direct regulation. This distribution is called an initial

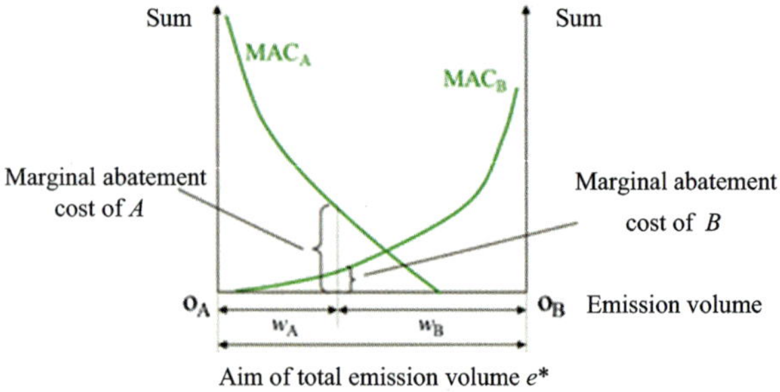

Fig. 15.16 Initial distribution of emission allowance

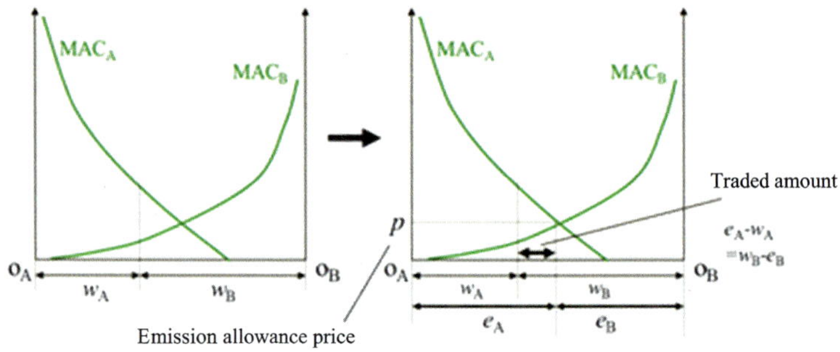

Fig. 15.17 Equilibrium in emission trade market

distribution. The emission trading market is formed by approving emission trading. The emission allowance price is decided according to the balance between supply and demand. The upper limit of emission is decided, and total volume control is conducted (assigned amount).

Figure 15.16 adds w_A and w_B, the initial distributions of *A* and *B*, respectively, originally shown in Fig. 15.12. The government distributes the desirable emission volume e^* to two companies, so the sum of w_A and w_B is e^*. The marginal abatement cost of *A* and *B* in the initial distribution is compared, showing that *A* is higher. This implies that the marginal cost equalization failed and also that this distribution is inefficient. At the same time, a trading opportunity exists. The emission allowance price is the fee a company pays when buying the allowance to increase one unit of emission from another company. From the discussion in Sect. 15.3, the marginal abatement cost of one company is the benefit gained when the company increases emissions by one unit. Therefore, it is more beneficial for the company when the emission allowance price is lower than the marginal abatement cost. Conversely, it is more beneficial for the company to sell when the emission allowance price is greater than the marginal abatement cost.

Currently, the marginal abatement cost of *A* is greater than that of *B* at initial distribution. When this happens, *A* is the consumer of emission trading and *B* is the supplier under a certain price, creating emission trading. This kind of opportunity

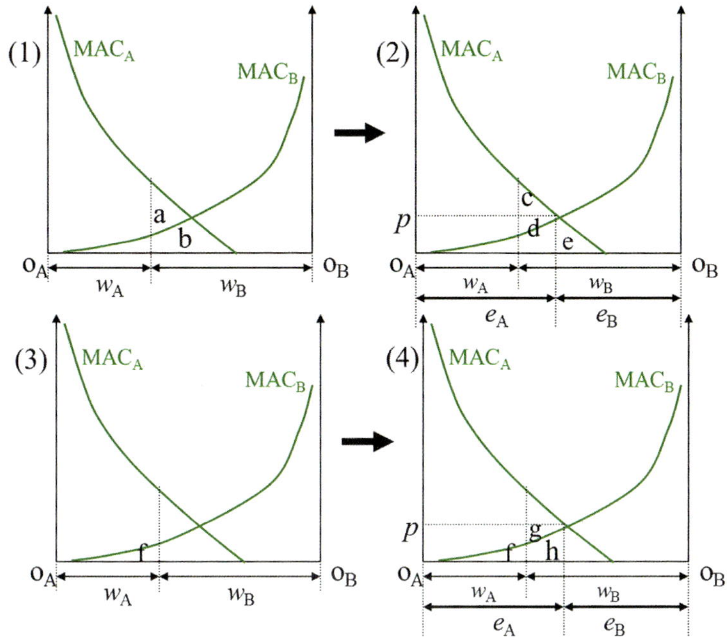

Fig. 15.18 Merit of emission trading

for trading exists as long as there is a difference in marginal abatement cost between them. In the market equilibrium, the emission volume that equalizes the marginal abatement cost of both companies becomes available. In other words, an optimum emission distribution is achieved via the market mechanism. This is shown in Fig. 15.17. Here, *A* purchases emission allowance of e_A-w_A(= w_B-e_B) from *B* at price *p*. Note that the equilibrium price *p* is the same as the marginal abatement cost for both companies.

Next, let's confirm that this trading is beneficial to both companies *A* and *B* in Fig. 15.18. When *A* makes distributed emission w_A without emission trading, the abatement cost of *A* is the area $a+b$ in Fig. 15.18 (1). When *A* buys emission allowance e_A-w_A from *B* and makes emission e_A, the payment needed to buy the emission allowance is the area *d* (rectangle) in Fig. 15.18 (2). The abatement cost is the area *e*. The sum of these is area *c* subtracted from $a+b$. This means *A* has gained a profit of area *c* from emission trading. On the other hand, the cost of *B* is the area *f* in Fig. 15.18 (3), without emission trading. The abatement cost after emission trading is $f+h$ in Fig. 15.18 (4), and the profit from selling the emission allowance is the area $g+h$ (i.e., the cost is $-(g+h)$). The sum is the area $f-g$. This means *B* gained a profit equivalent to the area *g* from emission trading.

The equilibrium explained above does not depend on the initial distribution. No matter how emission trading is distributed, the marginal cost of each company becomes equal to the emission allowance price, achieving equilibrium where the equi-marginal principle is established, as long as fair trading occurs. This is the merit of emission trading. However, we need to be aware that the profit of each company depends on its initial distribution. Emission trading distribution is equivalent to trading money. Therefore the profit of a company that receives a large share of the emission allowance will also be large. Thus, the decision of the initial distribution by the government is an issue of fairness.

One of the successful emission trading examples is that of sulfur dioxide in the United States in the 1990s. Another is the international carbon emission trading against global warming that was systematized in the EU in 2005. Other nations and areas are now considering introducing similar measures.

The trading system explained here is called the "cap and trade system." Another system is the "baseline and credit system," in which the effect of abatement in a pollution abatement project is approved. Examples include the Joint Implementation (JI) and the Clean Development Mechanism (CDM) stipulated in the Kyoto Protocol. These names were included under the blanket term of the Kyoto Mechanism, together with the international emission trading system actively in use now.

The emission trading system is not without problems. In addition to deciding on the total emission allowance and the initial distribution, there is a possibility that companies and nations may control the market control. In addition, costs are incurred in monitoring the system. If these problems can be overcome, emission trading can become an important environmental policy tool.

References

Hosoda E, Yokoyama A (2007) Environmental economics. Yuhikaku Publishing Co., Ltd., Tokyo, p 5

Ueta K (1996) Environmental economics. Iwanami Shoten Publishers, Tokyo, p 4

Ueta K, Oka T, Niizawa H (1997) Economics of environmental policies. Nippon Hyoron Sha Co., Ltd., Tokyo, p 24

Chapter 16
Environmental Policy

Shiro Hori

Abstract A variety of policies, direct regulation, economic measure, voluntary action, and information disclosure are applied in environmental problems. Optimal policies differ, depending on the contents of environmental problems and/or the circumstances of social systems. Economic measure, including both pricing policy and target policy, is theoretically effective policy; however, this policy works in condition of high price elasticity of demand and well-developed market. Compensation system for social damage is necessary when environmental damage happens. Policy analysis and assessment are required to assess policy. Tools such as cost-benefit analysis, cost-effectiveness analysis, and risk analysis are used for evaluating policies. However, these analyses can be used in assumption of the trust on the assessment by the general public. The discount rate has a great impact on climate change policy because cost and benefit generally occur at different times. A recycle-based society can be established based on the appropriate assessment of the costs and benefits of recycling and disposal.

Keywords Environmental policy • Direct regulation economic measure • Voluntary actions • Information dissemination • Policy mix • Compensation • Policy assessment • Cost benefit • Risk

16.1 Overview of Environmental Policy and Measures

16.1.1 Classification of Environmental Policies

There are a large number of policies applying in environmental problem, and each policy has its advantage and disadvantage. The environmental problems and the social system will decide which environmental policy should be adopted. Environmental policies include direct regulation, economic measure, voluntary actions, and information disclosure. These are policies to improve the environment; however it has to be noted that it is essential for ensuring the implementation of these policies to establish the pollution-preventing structure of companies which are the main

S. Hori (✉)
Fukuoka University, Fukuoka, Japan
e-mail: horishiro@adm.fukuoka-u.ac.jp

T. Shimaoka et al. (eds.), *Basic Studies in Environmental Knowledge, Technology, Evaluation, and Strategy*, DOI 10.1007/978-4-431-55819-4_16

Table 16.1 Classification of environmental policies

Means	Features	Incentive	Prerequisite environment	Case example
Direct regulation	Regulator standard, enforcement mechanism	Punishment, penalty	Well-established legal structure, implementation mechanism	Air pollution, water pollution, chemical substances
Economic method	Tax, grant, trading	Economic benefit	Well-developed markets	Emission tax, emissions trading
Voluntary actions	Voluntary statement, voluntary agreement	Social evaluation, avoidance of regulation	Social pressure, trust between society and companies	Pollution prevention agreement with local government, global warming prevention agreement
Information dissemination	Information disclosure	Social assessment	Social pressure, data development	PRTR

execution of environmental protection. Furthermore, damage compensation along with environmental pollution and destruction are also important in environmental policy systems.

Table 16.1 is a summary of each environmental policy. In order for effective implementation of each policy, the incentives and obligations are the key in each policy. The appropriate policy would be selected in accordance with incentive and cost. For example, incentives for business operators are needed to implement measures in avoiding penalties in direct regulation, gaining economic benefit in economic measure, avoiding restriction and improving social reputation in voluntary action, and avoiding risk in informational disclosure. Therefore, it is necessary to design a policy system whereby the incentives of these policies are appropriately exhibited.

16.1.2 Direct Regulation

In direct regulation, the government sets the regulatory standard, while it also monitors compliance situation. Therefore, direct regulation is, in other words, called as "command and control." The regulatory standard is the goal of a desirable environment which needs to be established. Then, emission standard and emission restriction rules are determined to achieve such a goal, thus placing emission restrictions on companies. The direct regulation is widely applicable to various environmental problems where regulatory standard can be scientifically and quantitatively established, for instance, recommended limit of toxins can be decided scientifically for the prevention of health hazard. The advantage of this policy is to make clear the goal of environmental target since the standard and the restriction rules are shared in relevant parties through public announcement.

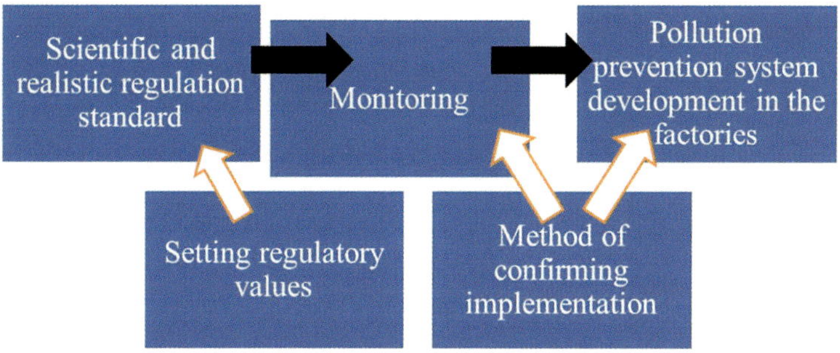

Fig. 16.1 Regulatory means

On the other hand, the government is expected to have information on which accurate regulatory standard can be established in order for this policy to be well enforced. If the government fails or mistakes to establish appropriate regulatory standard, the policy worsens cost performance or deterioration of other environmental situations. This issue is further explained in Sect. 16.4 "Assessment of Environmental Policies."

In addition, this system requires the pollution prevention structure and the monitoring by the government. The relationship between setting appropriate standards and monitoring by the government and system building by the practitioners is shown in Fig. 16.1.

The environmental policy in Japan is explained in the following.

First, the goal of a desirable environment, i.e., the Environmental Quality Standards, is established based on the Basic Environment Law. There are regulatory laws in each environmental field to achieve such an Environmental Quality Standards: the Water Pollution Control Law, Air Pollution Control Law, Noise Regulation Law, Offensive Odor Control Law, Soil Contamination Countermeasures Act, etc. Each law stipulates a set of regulatory standards for companies. To comply with regulations, companies need to establish departments and personnel in the corporation to take pollution-abatement actions. Details of the regulation are prescribed in the Pollution Prevention Managers Act. The Act identifies the environmental knowledge and the expertise with pollution prevention techniques which personnel in charge of environmental issues should have. It also requests that personnel in charge of environmental matters must be certified by the national qualification examination. Such system is very meaningful to develop human resources in charge of pollution abatement at companies to maintain the high level of skill. Recently, besides this legal system, voluntary environmental management standards, such as ISO14001, have been popular, and increasing numbers of companies are voluntarily carrying out system building and receiving certification. Both obligatory system building by the Manager Act and voluntary system building to take environmental certification work hand in hand.

16.1.3 Economic Method: When Are Economic Incentives Effective?

Economic incentives literally induce environmental improvement by using economic benefit. From the economic viewpoint, maintaining the environmental quality can be achieved by internalizing social cost (external cost) which is generated along with economic activities. In principle, companies must appropriately include the cost of pollution abatement in the production cost. If not, the government must take measures, such as subsidies and tax, to compensate social cost for realizing environmental quality in practice. Therefore, it is desirable that the ideal tax is set at the level of meeting the whole external cost (Pigovian tax). However, in reality, the calculation of such a tax level is extremely difficult, and in addition, the design of the tax system and collection system has similar difficulties. If the tax system is not designed to match social systems, the tax burden imposed on those in the social stratum may become disproportionate. This is why the whole notion of tax tends to have low social tolerance despite its theoretical efficiency. Therefore, taxation is implemented in the form of internalizing some part of the necessary external costs rather than covering the entire social cost. This aims the formation of a market that external costs are partly reflected on the product price. The environment is improved by reducing the supply of goods which have a negative influence on the environment.

Moreover, a cost of monitoring the market condition is incurred in order to create an optimal environment by economic measure. The effectiveness is heavily dependent on each market situation and its system design.

Cases of economic measure include the ETS (emissions trading system) of the EU, carbon tax, sulfur oxide allowance trading in the USA, emission trade in China, green tax in China and Taiwan., and taxation on fuel in Japan.

16.1.4 Voluntary Actions

Under the condition where the capability of government and the market system is undeveloped, it is necessary for companies to take voluntary actions. Voluntary actions by companies include a preemptive type where companies voluntarily carry them out, an agreement type where companies and government conclude agreements, and a voluntary program type where companies go along with a government guidance scheme.

Responsible Care, adopted by the chemical industry, is a global preemptive type of voluntary action by companies. Led by the International Council of Chemical Associations (ICCA), Responsible Care aims at reducing the environmental burden of chemical substances. The agreement type includes the agreement on pollution prevention between local governments and companies in Japan. This agreement was implemented prior to settlement of pollution regulations by the government and worked effectively in developing pollution-abatement measures with consideration for local circumstances. This type is effective as a measure against pollution problems in large cities, which are increasingly the area concentration of pollution. Major cities in China, where pollution became serious, conduct voluntary energy-saving measures led by local governments. Local governments conclude agreements with 100 selected companies as the 100-company project in local communities in addition to the 1,000-company project of the central government. Voluntary actions taken by companies can be seen worldwide, following schemes initiated by governments. Cases of program type include the greenhouse gas emissions reduction program in the USA and the action program to arrest global warming, waste reduction planning, and volatile organic compounds reduction plan in Japan. Large reductions have been achieved in each case, verifying the effectiveness of voluntary actions.

As for the incentive for companies to take voluntary action, social reputation is a strong incentive. The announcement effect is also considered to be the incentive for companies to induce voluntary actions. The announcement effect works in that the company gains positive social evaluation by taking an action and/or making a statement. In fact, it falls within public relations activity, and that means it is desirable to strengthen social appeal in order to increase this effect. Guidance by the government toward the establishment of voluntary actions has an announcement effect. In addition, a major announcement effect can be expected by mass media bestowal of awards to good companies. These methods can achieve a greater level of success by cooperation between government and business organizations. In this regard, the announcement effect works effectively in all policy method and does not apply only to voluntary actions.

16.1.5 Information Disclosure

Information disclosure is the policy to achieve environmental improvement by disclosing relevant information of the environment-related activities of companies. The history of this policy is parallel with the history of social responsibility. The increase of the awareness of corporate social responsibility has triggered the spread of this system. These days, companies are expected to take environment-friendly actions based on a corporate social responsibility in addition to complying with regulations. The most popular method in this policy is to make and disclose a company environmental report. Rating companies according to their environmental performances and disclosing the information are effective methods.

The systems that regularly disclose information on carcinogenic chemical substances (PRTR in Japan, TRI in the USA, and RTR in the EU) are typical government-led information disclosure systems. Four hundred sixty-two chemical substances are designated in Japan (Class 1 Designated Chemical Substances). Reduction is promoted by disclosing the amount of emission and transfer. The total amount of emission and transfer was 450,000 t (FY 2010). The incentive for reduction is to grasp the accurate amount of toxins by each company, and thus, there is a considerable feeling of social pressure when emission levels are high compared with other rival companies. Therefore, this system will have little effect in a society where such social pressure is not felt by the companies.

16.1.6 Policy Mix

Various policy measures are applied for various environmental problems. Each policy measure has its characteristic advantage, and there is no single mighty policy that is applicable to all problems. An optimal environmental policy is one that differs according to the timeframe and local situation. For example, policy against pollution that causes a health hazard requires fast-acting property and also reliability of standard rather than cost-effectiveness. Direct regulation is necessary in such cases. Cost-effective measures are sought for actions toward creating a more desirable environment even though it does not adversely affect health immediately. For example, there is continuous dispute to select appropriate policy measures for climate change. Cost-effectiveness must be considered for such problems as climate change where the damage is uncertain in near term, but considered in long-term view, and a tremendous level of cost is incurred, if all abatement cost is added up as a total. Therefore, policy mix among direct regulation, economic method, and voluntary actions is assertively used.

In addition, in many cases, multiple policies instead of single policy were adopted for more effective countermeasures. A combination of multiple policies tends to redouble their effect to bring about a better result. Policy against volatile organic compounds in Japan requires different methods from those in traditional direct regulation against toxins to voluntary actions. It is, in fact, a combination of regulation for large companies and voluntary actions for small companies. In addition to the uncertainty regarding the impact of volatile organic compounds on the environment, though they are used in all industries, their emission forms are diverse, and uniform regulation is considered to exhibit poor cost performance. Therefore, voluntary actions have the potential for effective result through a reliable guideline by the government in such cases. On the other hand, uniform regulation is considered to be effective for large companies due to the large volume of emissions and their high social responsibility. Therefore, a complex policy of regulation and voluntary actions is conducted. Measures against air pollution by sulfur oxide in the USA are a combination of two policies: emission regulation and emissions trading. Emission regulation improves the whole air quality, while a system of reduction by more economical measures is introduced by emissions trading with the upper limit of emission from the viewpoint of preventing problems concerning a wider area such as acid rain. Another example is a policy mix of climate change levy and voluntary actions in the UK. The climate change levy in the UK was introduced in 2010. The tax is reduced or exempted for companies which voluntarily reduce emissions for 10 years, promoting more voluntary action.

As has been seen so far, combinations of optimal policies can change depending on the goal of environmental policies.

Complementary policies are also effective. For example, financial assistance for companies would be even if direct regulation is the optimal policy. In Japan, low interest loans and tax deduction for introducing pollution-abatement equipment have been conducted along with the development of pollution control laws in the 1960s. These are effective means for policy implementation.

16.2 Effectiveness of Economic Measures: Theoretical Analysis

Economic measures are theoretically efficient policy to pursue the emission reduction. However, several conditions are prerequisite for their effective function. Such conditions are examined here.

First, let's consider the condition for an effective pricing policy, that is, taxation and subsidiary. The key is the quantitative change of demand responding to a change in price. Observe this in Fig. 16.2. When the marginal benefit curve is an MBb, the reduction effect brought about by tax is b-b′, while an MBa curve proves the emission reduction effect of a-a′. The MBa case is one for low price elasticity that is often observed in case of necessity goods. Pricing policy has limited effect in such cases. For example, it can be observed that energy taxation is effective in developed countries, but is not so effective in developing countries. This is because a certain amount of energy demand can be reduced among with a considerable degree of energy usage in developed countries, while major energy demand is nearly a daily use as necessity goods in developing countries. As a result, demand did not change even if price would change. Taxation has such problems of difficulty in setting levels where no damage is caused, regressive tax, and limited effect when there are no alternatives (Turner et al. 1993).

Next, consider the conditions where emissions trading works effectively. This requires a perfect market, that is, the market has free and open competition and the participating business operators are able to compete without any restriction.

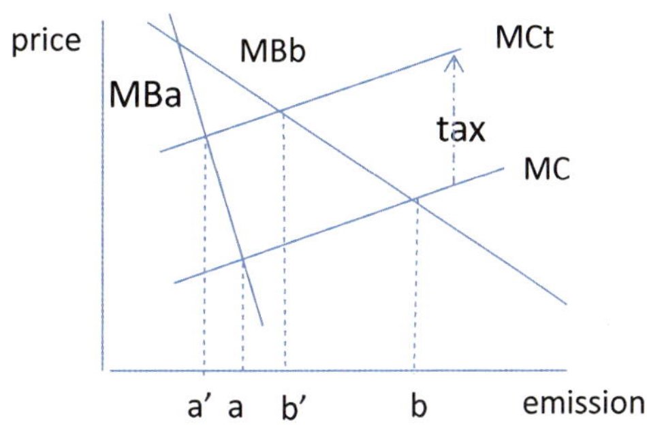

Fig. 16.2 Effect of pricing policy

However in the real world, markets are not perfect. Some markets are monopolized or oligopolistic by a small number of companies. Some markets have large companies who may enjoy a dominant position against other companies, and it may be able to control the prices. For this reason, policymakers are expected to intervene in markets to exclude dominance or secure transparency by distributing information in the market.

Now, let us consider which economic measure is more effective, pricing policy or quantitative target.

Wiseman compared these in a simple way. Under a condition where the effect of emission reduction is uncertain and the marginal cost curve is steeper than the marginal benefit curve of emission, quantitative target is more effective. In the opposite situation, tax policy is preferred when the marginal benefit curve is steeper than marginal cost curve. Observe this in Fig. 16.3. Consider a condition where reduction effect, that is, benefit, is uncertain. Suppose the regulatory agency misunderstand marginal benefit and acknowledged the false curve (MBfalse) as a preferred one instead of the true curve (MBtrue). In such a case, the regulatory agency would establish a false tax instead of a true tax. The tax rate which was correctly at c is moved to wrong point a; accordingly, the benefit area, abc, is lost, which is the social loss accompanying this misunderstanding. On the other hand, in quantitative target, d is acknowledged as the intersection point of MC and MB, instead of true point, c, resulting in moving the quantitative target from E to E′, and an additional cost area, cde, is created. In the upper part of Fig. 16.3, benefit area,

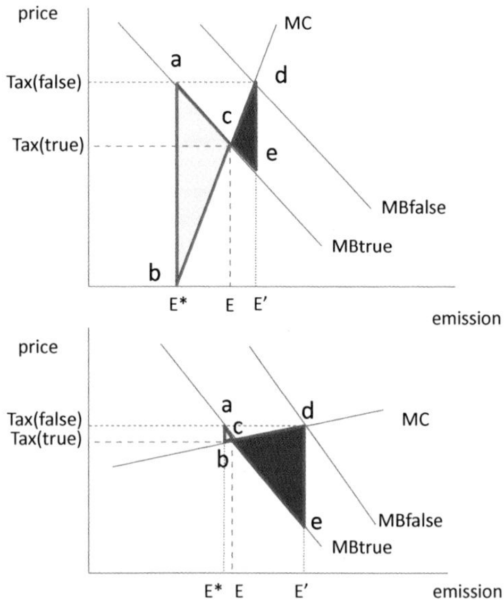

Fig. 16.3 Numerical goals and pricing policy

abc, is larger than loss area, cde; therefore, the quantitative target has less social loss than tax policy. The lower part of Fig. 16.3 shows the reverse case.

To compare pricing policy and target policy, let us examine two systems of renewable energy promotion, RPS (renewable portfolio standard, setting an obligatory amount of introduction and trading) and FIT (feed-in tariff, purchasing at fixed prices). In the RPS system, a target of renewable energy is set, and accordingly, electricity companies are obliged to introduce renewable energy. Renewable energy can be introduced over the whole electricity industry, so trading obligatory amounts among companies is acceptable for minimizing cost. On the other hand, FIT is a system where the purchasing price of renewable energy is set for each type of renewable energy to promote introduction.

A quantitative renewable energy target is set in the RPS system, so the amount to be introduced in the future can be predicted, but there is no guarantee of the purchase price for the generators that bring the financial risk in their business plans. On the other hand, there is a guarantee of purchasing at a fixed price in the FIT system, so it is easier for generators to build business plans, but there is a possibility of the cost remaining at a higher level than rational cost that burdens the consumers. Figure 16.4 shows how the cost reduction effect belongs to these two stakeholders. Here, suppose the marginal cost curve MC was changed to MC′ by price reduction. In such a case, the price is changed from P to P′ in RPS. The benefit accruing to generators is reduced by the area of PP’ZX. On the other hand, in FIT, the introduced amount is changed from Q to Q′ because P is fixed. As a result, the benefit accruing to generators is increased by XYP*. Meanwhile, the cost to consumers increases by the same amount.

There have been lots of discussions on the comparable effectiveness of RPS and FIT. For example, renewable energy is expected to increase by reduction of costs by technology development. Profit is distributed among generators in FIT, so FIT encourages more technology development. However, we can also say that RPS encourages technology development because it promotes competition. In either case, it should be noted that the superiority of one system varies, depending on the circumstances of the local community and business environment.

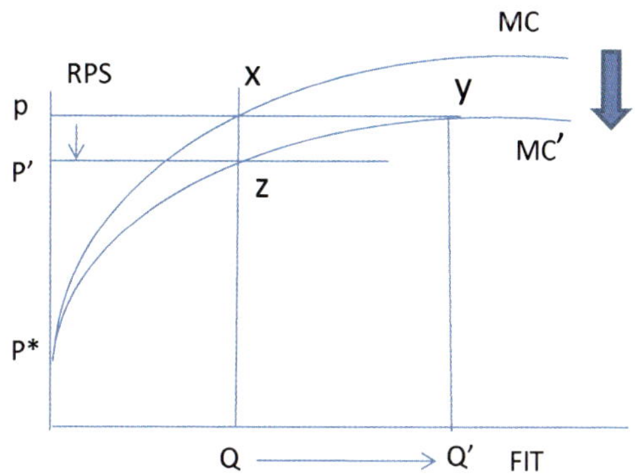

Fig. 16.4 Comparison between RPS and FIT (Source: Menanteau et al. (2003), modified)

16.3 Compensation Policy for Social Damage

As human activities and industrial production have expanded, damage to the environment and harm to people's health are inevitably brought out. A lot of policies are adopted to minimize such damage or harm; at the same time, we have to consider the policies to deal with compensation of social damage. The issue of compensation is who will be responsible and how to compensate the victims when they happened.

The Coase theorem teaches that compensation concerning damage to the environment is smoothly implemented by transactions of compensation money between polluters and victims. However, in reality, it is difficult to implement compensation by direct transaction between polluters and victims. This is because scientific evidences are necessary for establishing the source of the damage and a causal relationship between polluter and damage which is required for compensation responsibility, and a high amount of costs, namely, transaction cost, is inevitably incurred by victims. Coase theorem assumes that transaction cost is zero.

16.3.1 Legal Aspect of Compensation

A legal framework concerning damage compensation becomes necessary in order to decrease transaction cost and allow smooth implementation of compensation for victim. Pollution-Related Health Damage Compensation Act is established in Japan in 1973. The payment system of compensation by polluting companies (a financial burden on companies responsible for a wide range of air pollution is imposed as a levy) is established (Fig. 16.5). Aid for victims based on the Pollution-Related Health Damage Compensation Act is implemented in two categories. Category A is the case where the company causing the damage can be definitely specified, such as the case of Minamata disease (in this case, the polluter is Chisso Corporation) and itai-itai disease (in this case, the polluter is Mitsui Mining and Smelting Corporation). In this case, the polluter company primarily carries out their responsibilities of the compensation. The other is the case where the polluter companies are numerous, and a complex mix of pollutants from those companies causes social damage (category B). In this case, polluter companies discharge their responsibility through contribution to compensation fund.

Figure 16.5 illustrates a compensation scheme when payment is made by the polluters whose causal relationship to the victims is recognized. Next, let us see the case where health damage is actually happening and the victims have a dire need for aid, but there is no clear evidence of causal relationship between polluters and victim. In this case, in viewpoint of social welfare, aid fund was established jointly by governments and related companies. For example, prior to the establishment of the Pollution-Related Health Damage Compensation Act in 1973, the Act on Special Measures Concerning the Relief of Pollution-Related Patient was

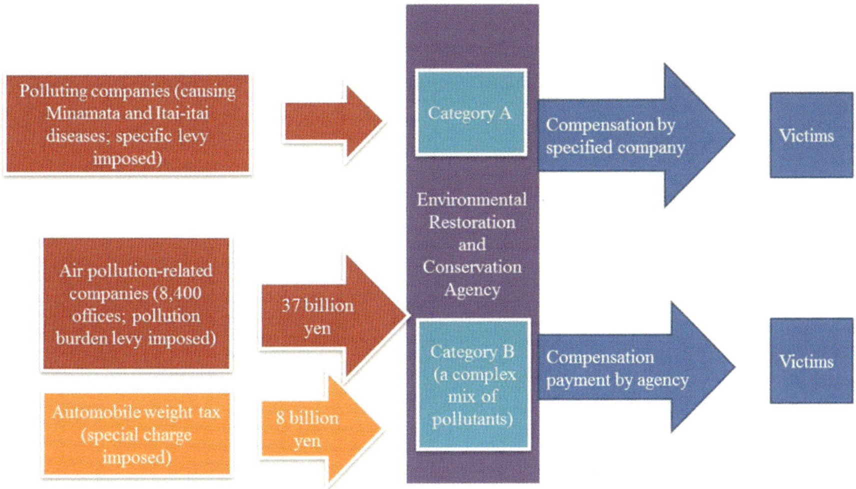

Fig. 16.5 Payment for pollution victims (scheme of the Pollution-Related Health Damage Compensation Act)

established in 1969 by funding from governmental contributions and business donation, providing aid for the victims. These aid systems are effective when related companies are numerous and a causal relationship is unclear. The case of Act on Asbestos Health Damage Relief falls in this category. Aid money is provided to the patients of respiratory diseases due to asbestos (victims of asbestos) as a whole based on this Act. This aid system was adopted because asbestos was used in a very large number of businesses sectors and it was difficult for the victims to make clear causal relationship with which occasion had affected their health.

The following case was pointed out by Yokemoto (2007) as an example of difficulty in applying the concept of polluters providing compensation. (1) The large amount of necessary relief cost is generated by continuing environmental pollution, exceeding the payment capability of the polluters, (2) expansion of industrial activity generated great many pollution-related sectors, and (3) cases of accumulated pollution where polluters cannot be specified. (1) is realized among the cases related to the Pollution-Related Health Damage Compensation Act. For example, Chisso Corporation which caused Minamata disease received assistance from the national government and local governments due to lack of company's payment capability. The same situation can be observed in the Tokyo Electric Power Company in the case of nuclear power plant accident. The compensation cost is over the payment capability of Tokyo Electric Power Co. Asbestos pollution case falls in the cases of (2) and (3).

At last, let us consider the range of responsible companies in charge of damage compensation. When a company directly causes harm to people and/or the environment, such company should bear responsibility (Polluter Pays Principle). However, the definition of polluters is becoming unclear. Let us consider environmental responsibility in automobile pollution. A direct source of automobile pollution is generated from cars. So any driver who drives a car is expected to be a polluter. However, compensation money cannot be collected from individual drivers. Therefore, part of the automobile weight tax paid by the owners of automobiles is allocated to the fund for pollution damage. What about the responsibility of other stakeholders besides drivers? Automobile pollution occurs around the busy road. The liability for concentrating automobile traffic in the specific road is recognized to be with those entities which build the road (national government, local governments, and highway companies). The social responsibility of automobile manufacturers was also recognized. As is seen here, the range of responsibility of companies has been wider in recent years. This indicates that corporate social responsibility in addition to the legal responsibility is imposed as company's responsibility in the environmental damage.

16.3.2 Economic Aspect of Compensation

Compensation for social damage was explained as internalizing social cost accompanying economic activities in economic theory. Coase explained that environmental damage can be compensated through direct transaction between victims and polluters; however, it is difficult to transact social cost from victims to polluter, due to the difficulty of evaluating social cost. Therefore, victims are generated when such social cost is burden on the part of certain people without internalizing cost. Thus, Kapp said evaluation of social cost of damage was outside the reach of scientific research and responsible for political decision.

16.4 Assessment of Environmental Policies

In the previous chapter, it was explained that there are many policy measures to deal with environmental problems. Then, what criteria shall be applied to choose optimal environmental policies? Policy analysis and assessment method to define the criteria have been developed. Nowadays, analysis and assessment of policies are widely used in an evaluation of government policy. For example, it is mandatory in Japan to conduct policy assessment to evaluate the effectiveness of each policy in all the policies to be introduced. In the USA, assessment of effectiveness of any regulation is requested before introducing a new regulation.

There are numerous methods in policy analysis and assessment. Among them, we will start with cost-benefit analysis first. Here, the cost and benefit of designated policy are compared, and accordingly, the effectiveness of the policy is evaluated. Second, cost-effectiveness analysis is the method to assess the effectiveness of each policy, and thereby the priority of policy set for a certain problem is assessed. Third, risk analysis is used to assess the environmental damage by concept of risks. The risks for the environment are the uncertainty of a problem caused by pollution. These scientific assessment tools are inevitable to analyze policy effect.

16.4.1 Cost-Benefit Analysis: Case Study of the Policy

Cost and benefit must be considered when introducing environmental regulation. Let's start by the definition of cost and benefit. Cost can be described as compensation cost when environmental destruction is expected or occurred. It can also be described as an alternative cost to endure the environmental deterioration. This is equal to WTA (willingness to accept). Benefit is the value that people gain by maintaining or improving the environment. This is the total of value what individuals consider, so it is a total of payment that individuals are willing to pay. This is shown as WTP (willingness to pay).

Cost-benefit analysis can be thought of as a comparison of cost and benefit. Then, in which policy situation cost-benefit analysis can be applied effectively? The application of policy assessment upon introducing regulations was stipulated by the United States Executive Order in 1981, and it became mandatory to prove that the benefit would surpass cost by introduction of new regulation. This Executive Order was succeeded in the Executive Order in 1993. In the EU, EU Directive in 1996 prescribed cost-benefit analysis of regulations upon introducing air pollution regulations.

However, in fact, it is not easy to assess benefit to the environment. Various problems have been pointed out when such assessment is applied in environmental policies (Kolstad 2011). For example, a question is brought up if WTP and WTA can be assessed accurately in a situation with people only having access to limited information. This problem particularly becomes obvious in case to assess value of nature and biodiversity, or environmental destruction risk, as well as the value of human health. Due to the difficulty in quantifying benefit, assessment measures were changed from quantitative comparison to confirmation of gaining legitimate benefit in the Executive Order in 1993. Furthermore, when those who bear the cost and those who receive the benefit are different, there is a question of how to gain consensus between them. In addition, there is a time problem when there is a time discrepancy between beneficiary and victims. Such a gap in time can occur in major environmental problems. Introducing a discount rate is a tool that will provide an answer. The discount rate will have a key role on evaluation cost and benefit. In other words, cost-benefit ratio can change according to the discount rate.

The discount rate is explained, using formulas (16.1) and (16.2). When there is no discount rate, the ratio of cost and benefit is shown as C/B. Suppose there is a gap of t years between benefit and cost, and annual interest rate is r, the current benefit will take a form of (16.1) in t years. Here, the value in t years is modified to the current value of (16.2).

$$B(1+r)^t \tag{16.1}$$

$$B/(1+r)^t \tag{16.2}$$

How can the discount rate affect the evaluation of problems?

This can be shown by a simple example. For example, ask the question of which is preferred, receiving 1,000 yen today or in 1 year. Most of the people will answer "today." It is because value will decrease in 1 year by the portion of the interest rate. Then what if 1,100 yen can be received in 1 year? What about 1,200 yen? When the amount of receiving money in 1 year is raised, gradually, more and more people will want to receive in 1 year. Those who will be satisfied with 1,100 yen in 1 year understand the discount rate to be 10 %, while those who will be satisfied with 1,200 yen understand the discount rate to be 20 %. That is to say, individual action can change depending on the discount rate which leads to different future values. Whatever the case, it should be noted that overestimation of the discount rate will impose a burden of cost on the future generation (Turner et al. 1993).

16.4.2 Cost-Effectiveness Analysis: Comparison Between Multiple Policies

Cost-effectiveness analysis is used for prioritizing policies. Cost-effectiveness analysis is useful when considering policy options.

Needless to say, the policy measure with the lowest cost should be selected in order to achieve a goal economically. Let's consider the case of carcinogenic substance regulation as this analysis. In this case, the mortality rate of cancer is considered as a cost. The risk of developing cancer is shown as risk of death per capita. The risk is calculated by how much the risk of cancer develops by the exposure to toxins. Additionally, the cost for implementing a measure which can reduce the risks is calculated, and the ratio of cost-effectiveness is calculated. Table 16.2 shows a comparison of the cost-effectiveness of various measures taken in the USA.

Table 16.2 Cost-effectiveness analysis

	Deaths per 1 million people exposed	Cost to avoid 1 death (million USD)
Trihalomethane in drinking water	420	0.2
Benzene fugitive emissions	1470	3.4
Asbestos occupational exposure	3015	8.3
Benzene occupational exposure	39,600	8.9

Source: The Council on Environmental Quality (1991), Turner et al. (1993) modified

16.4.3 Risk Analysis

Risk analysis is a tool to quantitatively measure the predicted impact of a certain environmental problem. The end of environmental policy is improving environmental problems to secure safety for people's health and environmental preservation. Quantitative analyses are necessary to set the goals of safety and environmental preservation. Risk analysis compares and sets priority among relative safety and environmental issues. Uncertainty exists in assessing environmental policy because the environment belongs to the whole nature system. What we can say is that risk analysis measures the degree of this uncertainty. This is represented by the degree of uncertainty of the impact (damage) on the environment and probability of exposure. Without accuracy in this risk analysis, accurate evaluation of environmental policy may not be conducted.

This exact example of risk analysis was applied to set the emission standard for benzene in Japan. The carcinogenicity in benzene is a confirmed fact, but it is not an acutely toxic substance with a threshold, and it is widely used under rigorous regulations. The regulatory value of benzene was estimated as a mortality risk of 10^{-6}. This is the risk of carcinogenicity, and regulation based on this method has been applied to numerous substances ever since.

As is explained here, accurate risk assessment enables choosing an optimal policy with minimum risk. On the other hand, unsound risk assessment may rather lead to environmental deterioration. There was such a case in Peru (Nakanishi 2010). Chlorine sterilization in tap water is known to cause trihalomethane, a known carcinogen. Therefore, the US EPA suspended chlorine sterilization in tap water. The Peruvian government followed the example and ran a similar policy. As a result, Peruvian tap water went without sterilization, which contributed to an outbreak of cholera. It is said that this cholera epidemic killed approx. 7,000. This incident came about by not comparing the carcinogenic risk associated with trihalomethane and the mortality risk of a cholera epidemic. This happened in Peru, but similar incidents have been happening all over the world. The use of ethylene dibromides (EDBs) as a disinfectant is banned in the USA, and its carcinogenic risk was 0.0004 %, while the carcinogenic risk of molds on grain is usually 0.03 %, which can go as high as 1 % in an environment where mold is prone to growth. This poses the question of which is more beneficial for human health – banning EDBs or allowing its use? (Nakanishi 2010).

Next, let us see a case of risk analysis in Japan. Measures against dioxin in Japan began when dioxin was detected in the soil around incinerators. As a result, measures against dioxin progressed focusing mainly on measures for incinerators, that is, by raising the temperature of incineration. Of course, the measure for incinerators is highly effective to reduce dioxin exposure. Measures against dioxin at incinerators are divided into two stages, namely, first measures (emergency measures) and second measures (permanent measures). The cost of extending life expectancy by 1 year in emergency measures is 7.9 million yen/capita/year, while that in permanent measures is 150 million yen. We can see that the cost-effectiveness of permanent measures is very much lower than that of emergency measures.

16.4.4 Risk Communication

Cost-benefit analysis is used for selecting policy. So it is a social decision tool making a regulatory framework by quantitative assessment as well as a scientific policy assessment. However, cost-benefit analysis, cost-effectiveness analysis, and risk analysis which we have discussed about are not complete by themselves as a social decision tool. Even if such scientific assessment method plays a key role, the trust of the general public in those assessments is prerequisite. In the real world, the amount of information, people have access to, is limited, and as a result, there is a gap between scientific assessment (objective risk) and public awareness (subjective risk) (Fig. 16.6). Kahneman showed in his study of psychology that the awareness and behavior of people are different; that is also different from what is based on rational theory. For example, risk awareness of people tends to assess continuous risks into a lower risk category but large-scale risks into a higher risk category. Risk management is required to fill up such a gap between public awareness and scientific assessment. Accurate analysis and media for transmission of information are the two pillars of risk management. The information dissemination is called risk communication. Without risk communication, accurate risk information is not shared, and accurate policy is not made.

When this trust is lacking, what kind of problem will happen in policy? The safety of mad cow disease is a case where the information announced by researchers and government suffered a loss of credibility. In this case, the

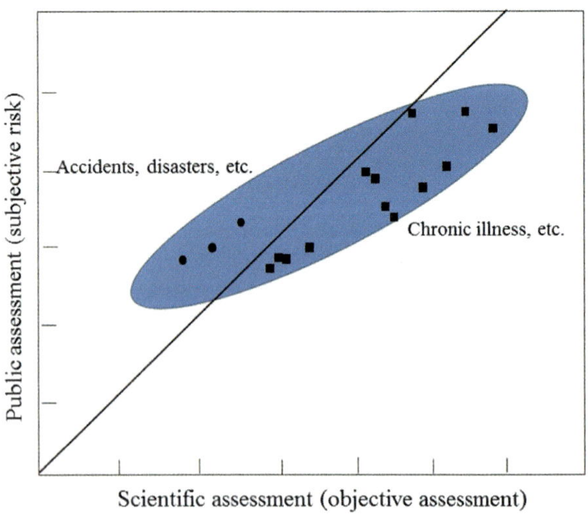

Fig. 16.6 Objective risk and subjective risk (Source: Slovic et al. (1979), Kolstad (2011) modified)

credibility of policies is lost by the delay in appropriate measures and information disclosure, resulting in a demand for unnecessary measures beyond the appropriate risk level.

16.5 Application of Policy Assessment for Some Problems

Policy assessment is useful for drawing up the prescription for the various environmental problems. However, some environmental problems require additional point of attention. Global environmental problems and establishment of a recycle-based society are the most common environmental issues but are hard to deal with. Let's examine those here for the example of policy assessment.

16.5.1 Global Environmental Problems

Some environmental issues are given attention on a global scale. The ozone layer destruction problem and the climate change problem are the two biggest problems in the viewpoint of global issues. All countries must cooperate to tackle global environmental problems where no global government exists. However, there is difficulty for countries to tackle the problem together. In the ozone problem, the victims (beneficiary) and cost-paying countries are almost the same. On the other hand in the latter problem, the victims and payers are different, which makes the climate change problem more difficult to reach appropriate policy agreement. In addition, cost-effectiveness in the ozone layer problem is 1:11, while it is 1:0.5 in the climate change problem (Nordhaus and Boyer 2000). We can see that the solution for the ozone layer destruction problem is going smoothly, while the solution for the climate change problem is anything but simple.

There is another difficult issue in the climate change problem; that is the gap between the generation bearing the cost of measures and the generation receiving the benefit. Traditional cost-effectiveness analysis is made on the premise that cost and benefit occur simultaneously. However, global environmental problems are not so easy. In particular, the climate change problem will require 100–300 years to stabilize the concentration of carbon dioxide, several hundred years to stabilize the temperature, and another several hundred years at least to stabilize the sea level. Even if the current generation takes appropriate measures, it will take a very long time for their effects to appear. Therefore, we cannot compare the measures (cost) and profit (benefit) unless we convert them at the same point in time. The discount rate is required to deal with the issue.

The discount rate brings about major differences in ideas for climate change measures. Sir Stern analyzed policy of climate change at the request of the UK government and compiled the report, so-called Stern Review (Stern 2006). According to the conclusion of this Stern Review, early actions for climate change lower the cost and increase the benefit. Stern assessed the cost-benefit ratio at 1:10. This figure is much higher than the one suggested by Nordhaus. Why such a difference? Stern uses a discount rate of 0.001, while Nordhaus uses the rather hefty 0.015. That is, the difference between the current value and future value is smaller in the figure of Stern than the one of Nordhaus. This indicates that the future value is higher evaluated in Stern Review. As a result, Stern concluded that emission control would rise to 53 % by 2015, while Nordhaus concluded that it would rise to merely 15 %.

Another important point in a dispute of climate change problem is the fairness of cost burden among countries. Let's consider this issue by viewpoint of nature of climate change. The climate change problem was prescribed by the United Nations Framework Convention on Climate Change in 1992 and the Kyoto Protocol in 1997. Constructing a new framework of climate change targeting at 2015 was agreed on at COP17 in Durban in 2011, and it was decided at COP19 in Warsaw in 2013 that new framework includes three points; namely, all the nations should join; each nation should voluntarily make their contributions toward reduction of GHG; and a review scheme should be established to confirm the fairness and accuracy of contributions. Policy and measure of reduction activity in developing countries is one of the controversial issues in the next framework. Developing countries are exempted from obligation in the Kyoto Protocol, but are requested to join in the next framework. The comparison of the marginal abatement cost of measures in various countries shows us that the cost-effectiveness of measures is higher in developing countries than in developed countries. Therefore, taking more measures in developing countries is economically rational when only the marginal abatement cost is considered. However, the priority of actions for the climate change is not solely judged by the marginal abatement cost. There is a viewpoint that social damage by climate change is caused by accumulated GHG emissions in the past. If this idea is adopted, the cost of measures in each country should be a fair burden in accordance with their past contribution to pollution, not merely cost-effectiveness of current measures. This is a complex case of how to allocate the responsibility when an accumulated environmental damage occurs (or is expected to occur).

16.5.2 Waste and Recycling

Next, let's consider the waste and recycling issue. Waste and recycling seem to be the local issues; however, waste is the common problem occurring all over the world. In addition, waste is transferred beyond the border. So, reducing waste and promoting recycling is one of the major issues in the global environmental context. Recycling issue became serious worldwide that the economic development as citizen life becomes affluent with home products and the large amount of waste fill in the society. It is desirable to reduce the amount of waste and promote recycling toward a recycle-based society. In Japan, waste is divided into two categories: general waste consisting of waste produced from households and small and middle enterprises that is collected and treated by local municipalities and industrial waste which is produced and treated by the big companies. The amount of general waste is 43.39 million tons (2011) with a recycling rate of 20 %, while industrial waste amounts to 385.99 million tons with a recycling rate of 53 %. Industrial waste has a higher amount and higher recycling rate than general waste. This is because general waste is a mixture of diverse kinds of waste, which make separation for recycling difficult.

Regarding to a policy measures toward a recycle-based society, various policies are examined. By law, a certain rate of recycling in waste is set for obligation of big companies producing a large volume of waste. There are several specific laws regulating certain sectors: recycling for home appliances by the Law for the Recycling of Specified Kinds of Home Appliances, automobiles by the Law for the Recycling of End-of-Life Vehicles, waste generated during construction by the Construction Material Recycling Law, and waste food by the Law for Promotion of Recycling and Related Activities for Treatment of Cyclical Food Resources.

In the case of general waste where a great many parties are involved in producing waste, there is difficulty in enforcing people by regulatory measures. Instead, economic measure is more applicable. When the recycling cost becomes lower than the incineration cost, it should encourage recycling. However, in reality, incineration cost is lower; as a result, the recycling rate remains low.

The reason for the slow spreading of general waste recycling is attributed to the fact that collection and treatment of general waste are considered to be the responsibility of local governments where the market mechanism is distorted. As a result, the incineration fee charged by municipalities is far lower than the recycling fee. In other words, official subsidiary is provided for incineration services which discourages recycling. It means that the treatment fee at incinerators does not reflect the actual cost, and municipalities offer incineration by fees not covering the cost. In the future, such a situation will need to be improved so that recycling should be handled according to the market mechanism.

Furthermore, it is pointed out that benefit is not accurately assessed in cost-benefit assessment of recycling. Recycling does not only reduce the incineration cost but also offers a wide range of benefits such as saving natural resources by recycling, reducing the environmental burden, and returning profit to the local community by the "locally produced, locally consumed" system. Such an assessment of policy including the value of social benefit is not well conducted. The lack of assessing social benefit is one of the reasons for the underperformance of recycling.

References

Kolstad C (2011) Environmental economics, 2nd edn. Oxford University Press, New York.
Menanteau P, Finon D, Lamy M (2003) Prices versus quantities: choosing policies for promoting the development of renewable energy. Energy Policy 31:799–812
Nakanishi J (2010) Risk studies of food. Nippon Hyoron Sha Co., Ltd, Tokyo.
Nordhaus WD, Boyer J (2000) Warming the world: economic models of global warming. MIT Press, Cambridge, MA
Slovic P, Fischoff B, Lichtenstein S (1979) Rating the risks. Environment 21(14–20):36–39
Stern N (2006) The economics of climate change: the stern review. Cambridge University Press. www.hm-treasury.gov.uk/stern_review_report.htm
The Council on Environmental Quality (1991) Environmental quality: 21st annual report 1990. US Government Printing Office, Washington, DC
Turner RK, Pearce D, Bateman I (1993) Environmental economics, an elementary introduction. Johns Hopkins University Press, Baltimore
Yokemoto M (2007) Responsibility and compensation of environmental damage and cost bearing. Yuhikaku Publishing Co., Ltd, Tokyo.

Chapter 17
Quantitative Analysis Method of Environmental Burden Using Input-Output Models

Ryoji Hasegawa

Abstract As a prerequisite to fully consider the socioeconomic situation in one country at a certain time and to implement appropriate environmental policies there, methodology and analysis models need to be established to quantitatively measure the environmental burden generated by human activities. This chapter outlines the use of input-output models for environmental analysis and case studies, as one method of quantitative analysis of environmental burden. First, the essence of an input-output table is explained, followed by an explanation of the basic framework of an input-output model and its application for environmental analysis. In addition, the emission structure of the environmental burden in Asia is examined by introducing international case studies of input-output models for environmental analysis. Due to limited space, the explanation of the input-output theory is given in simplified form in order to comprehend the general framework and purpose of environmental analysis by an input-output model. Please refer to Miller and Blair (Input-output analysis: foundation and extensions, 2nd edn. Prentice-Hall, Englewood Cliffs, 2009) and Shishido et al. (Input-output analysis handbook. Toyo Keizai Inc, 2010), for example, in order to fully understand the basic principles and application of input-output theory.

Keywords Input-output table • Leontief inverse • Induced CO_2 emissions • Carbon leakage

17.1 Introduction

As a prerequisite to fully consider the socioeconomic situation in one country at a certain time and to implement appropriate environmental policies there, methodology and analysis models need to be established to quantitatively measure the environmental burden generated by human activities. This chapter outlines the use of input-output models for environmental analysis and case studies, as one

R. Hasegawa (✉)
Faculty of Global Business, Osaka International University, Moriguchi, Japan
e-mail: hasegawa@oiu.jp

T. Shimaoka et al. (eds.), *Basic Studies in Environmental Knowledge, Technology, Evaluation, and Strategy*, DOI 10.1007/978-4-431-55819-4_17

method of quantitative analysis of environmental burden. First, the essence of an input-output table is explained, followed by an explanation of the basic framework of an input-output model and its application for environmental analysis. In addition, the emission structure of the environmental burden in Asia is examined by introducing international case studies of input-output models for environmental analysis. Due to limited space, the explanation of the input-output theory is given in simplified form in order to comprehend the general framework and purpose of environmental analysis by an input-output model. Please refer to Miller and Blair (2009) and Shishido et al. (2010), for example, in order to fully understand the basic principles and application of input-output theory.

17.1.1 What Is an Input-Output Table?

An input-output table is a kind of flowchart originally devised by Wassily W. Leontief, showing the transactions of one economy (country, regions, etc.) over a certain time period (usually 12 months). In more specific terms, an input-output table shows economic activity in detail at a certain time, especially from the viewpoint of transactions between industries. It does this by entering transactions of goods and services between various sectors, including industry, household, and government, into one table.

A System of National Accounts (SNA) is one of the economic statistical systems describing the quantity and structure of flow transactions in the economy of a country. It consists of five parts: "national income accounts," "input-output table," "flow of funds accounts," "balance of payments table," and "national balance sheet." For the purpose of rudimentary understanding, an input-output table can be regarded as an economic statistical table showing the complete state of an economy.

Next, in order to understand an input-output table in a practical context, it is necessary to understand how to read it, as well as understand its structure as an economic statistical table. Figure 17.1 is an outline table to show the structure of an input-output table. One of its key features is that the production value (value of sales) of each industry is entered from two viewpoints. The market structure of the production value (output) of each industry can be read from the rows (horizontal) of transaction amount in the table, while the cost structure of the production value (input) of each industry can be read from the columns (vertical) of transaction amount. For example, focus on manufacturing in Fig. 17.1. What we can read is that manufacturing sells products worth 3 billion yen to agriculture, 12 billion yen to manufacturing (itself), 10 billion yen to commerce, 10 billion yen to consumers, and 15 billion yen as investment assets, and the total output amounts to 50 billion yen. At the same time, when the cost is broken down, we can read that this sales amount of 50 billion consists of intermediate goods input worth 3 billion yen, 12 billion yen, and 12 billion yen from agriculture, manufacturing, and commerce sections, respectively, 15 billion yen of labor cost, and 8 billion yen of profit.

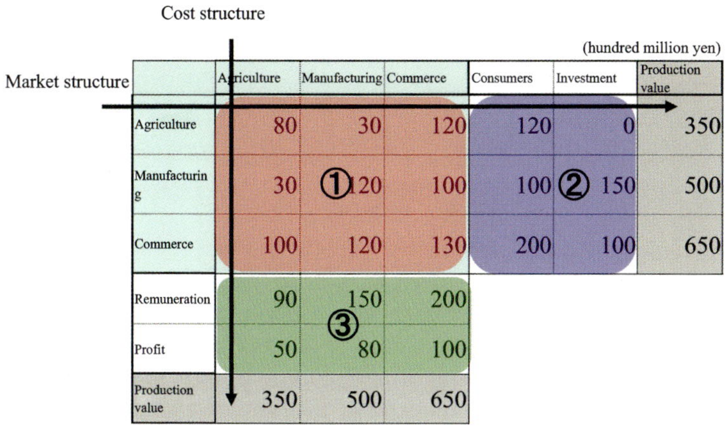

Fig. 17.1 Outline figure of input-output table

As is seen here, unlike national income statistics, the transactions of intermediate goods input among industries can be understood from an input-output table by simultaneously showing the structure of the transaction amount of each industry in two dimensions. Item (1) in Fig. 17.1 is called the endogenous sector, (2) the final demand sector, and (3) the value-added sector. In an input-output model, (1) the endogenous sector is literally handled endogenously. Then (2), the final demand sector and (3) the value-added sector are handled exogenously. Therefore, these are put together and called the exogenous sector.

Next is the explanation of an input-output model. Information is given on a quantitative model which has the most application examples. Table 17.1 is a mathematical representation of an input-output table using numerical formulae. For simplification, it has fewer sectors than Fig. 17.1, and import/export (overseas transactions) are entered. The balance formula is structured in the row direction in quantitative models. Therefore, the following quantitative model is developed as seen below in Table 17.1 and the following equations:

$$\begin{pmatrix} x_1 \\ x_2 \end{pmatrix} - \begin{pmatrix} a_{11}x_1 & a_{12}x_2 \\ a_{21}x_1 & a_{22}x_2 \end{pmatrix} + \begin{pmatrix} f_1 \\ f_2 \end{pmatrix} + \begin{pmatrix} e_1 \\ e_2 \end{pmatrix} - \begin{pmatrix} m_1 \\ m_2 \end{pmatrix}$$

$$\begin{aligned} &\Leftrightarrow \mathbf{x} = \mathbf{A}\mathbf{x} + \mathbf{f} + \mathbf{e} - \mathbf{m} \\ &\Leftrightarrow (\mathbf{I} - \mathbf{A})\mathbf{x} = \mathbf{f} + \mathbf{e} - \mathbf{m} \\ &\Leftrightarrow \mathbf{x} = (\mathbf{I} - \mathbf{A})^{-1}(\mathbf{f} + \mathbf{e} - \mathbf{m}) \end{aligned} \tag{17.1}$$

where $\mathbf{x}$ in Formula (17.1) indicates the output vector, $\mathbf{A}$ is the input coefficient matrix, $\mathbf{f}$ is the final demand vector, $\mathbf{e}$ is the export vector, $\mathbf{m}$ is the import vector, and $\mathbf{I}$ indicates the identity matrix. Each factor in the input coefficient matrix $\mathbf{A}$ is a value, that is, each transaction amount in the endogenous sector, divided by the input of each industrial sector, the result indicating the percentage of each intermediate goods input value in the input. Therefore, each factor in the endogenous

Table 17.1 Mathematical presentation of input-output table

	First sector	Second sector	Final demand	Export	Import	Production value (Output)
First sector	$a_{11}x_1$	$a_{12}x_2$	f_1	e_1	m_1	x_1
Second sector	$a_{21}x_1$	$a_{22}x_2$	f_2	e_2	m_2	x_2
Value-added	v_1	v_2				
Production value (Input)	x_1	x_2				

sector, that is, a transaction of intermediate goods between industries, can be described as $a_{ij}x_i$ as shown in Table 17.1. When **f**, **e**, and **m** are exogenously given in Formula 17.1, the output **x** is endogenously determined depending on $(\mathbf{I}-\mathbf{A})^{-1}$. $(\mathbf{I}-\mathbf{A})-^{1}$ is called the Leontief inverse and can be interpreted as the scale of the ultimate economic repercussion multiplier generated in the total economy when final demand is produced.

Formula 17.1 is called an import exogenous model since imports are treated exogenously. In such an import exogenous model, a balance formula is established by deducting imported goods included in intermediate goods and final demand, by lumping together. Therefore, the input coefficient matrix (**A**) contains imported goods, and the production value (induced production value) determined by the $(\mathbf{I}-\mathbf{A})^{-1}$ type of Leontief inverse includes overseas production value generated for producing imported goods. In such circumstances, domestic production value is calculated on the assumption that the imported goods were produced in Japan as well.

On the other hand, in the import endogenous model where imports are treated endogenously, there is a presupposition that the demand for imported goods depends on the scale of domestic demand such as intermediate goods and final demand. Therefore, we assume that the demand for imported goods is at a certain ratio of domestic demand.

In such circumstances, import vector (**m**) is modified as shown below:

$$\mathbf{m} = \begin{pmatrix} \hat{m}_1 & 0 \\ 0 & \hat{m}_2 \end{pmatrix} (\mathbf{Ax} + \mathbf{f}) \tag{17.2}$$

where (**Ax** + **f**) in Formula 17.2 indicates domestic demand, and it is considered that the demand for imported goods is at a certain ratio ($\hat{\mathbf{m}}$) of domestic demand; $\hat{\mathbf{m}}$ is called the import coefficient vector and is derived by $\hat{m}_i = \frac{m_i}{a_{ij}x_i + f_i}$ when each element is $\hat{m}_i$.

The import endogenous model is derived by modifying Formula 17.1 as shown below:

$$\begin{aligned}&\mathbf{x}=\mathbf{A}\mathbf{x}+\mathbf{f}+\mathbf{e}-\mathbf{M}(\mathbf{A}\mathbf{x}+\mathbf{f})\\&\Leftrightarrow \mathbf{x}-\mathbf{A}\mathbf{x}+\mathbf{M}\mathbf{A}\mathbf{x}=\mathbf{f}+\mathbf{e}-\mathbf{M}\mathbf{f}\\&\Leftrightarrow \{\mathbf{I}-(\mathbf{I}-\mathbf{M})\mathbf{A}\}\mathbf{x}=(\mathbf{I}-\mathbf{M})\mathbf{f}+\mathbf{e}\\&\Leftrightarrow \mathbf{x}=\{\mathbf{I}-(\mathbf{I}-\mathbf{M})\mathbf{A}\}^{-1}\{(\mathbf{I}-\mathbf{M})\mathbf{f}+\mathbf{e}\}\end{aligned}\tag{17.3}$$

Formula 17.3 is the import endogenous model, and $\mathbf{M}$ in the formula indicates a square matrix diagonalizing import coefficient vector ($\hat{\mathbf{m}}$). The Leontief inverse is shown as $\{\mathbf{I}-(\mathbf{I}-\mathbf{M})\mathbf{A}\}^{-1}$ in the import endogenous model. In such circumstances, the induced production value based on the production of imported goods is deducted, and only the domestically generated production value is included. Needless to say, the scale of the economic repercussion multiplier is smaller in the $\{\mathbf{I}-(\mathbf{I}-\mathbf{M})\mathbf{A}\}^{-1}$ type of Leontief inverse than in the $(\mathbf{I}-\mathbf{A})^{-1}$ type.

17.1.2 Environmental Analysis by Input-Output Method

In addition to monetary transactions between each sector which an ordinary input-output table deals with as the subject of entry, the table can be applied to environmental analysis of input-output theory by including environmental burden and resources/energy demand, that is, generated along with such economic activities, in the input-output model. Generation of environmental burden accompanying economic activities is strongly related to the complex transaction structure existing between each sector of industry, household, and government. One type of economic activity may have an impact on generation of environmental burden either directly or indirectly between sectors through the economic ripple effect, which makes it difficult to clearly understand the causal relationship between generation of environmental burden and economic activity between sectors. However, it makes it possible to comprehensively and systematically measure the generation of environmental burden, which has a complex causal relationship with economic activities detectable by applying input-output analysis to enable quantitative understanding of the economic ripple effect.

The input-output model for environmental analysis is established at an advanced level, and there are a large number of case studies. Environmental analysis by the input-output method can be divided into two approaches.

The first approach involves making a supporting table concerning environmental burden. A supporting table shows the input and output amount of employment and materials in addition to the basic transaction table, where monetary transaction flow is entered such as in Fig. 17.1. A supporting table of various types of environmental burden enables environmental analysis by drawing out a generating unit of environmental burden and entering it into the input-output model formula. As for a publicly available supporting table of environmental burden, *Embodied Energy and Emission Intensity Data for Japan Using Input-Output Tables (3EID)* by Nansai and Moriguchi (2012) published by the National Institute for Environmental Studies is well known as an industry-classified database concerning energy consumption and environmental burden caused by it in Japan.

The second approach involves expanding the basic transaction table of an input-output table and applying it to environmental analysis by entering emission of environmental burden as outputs and by establishing an industry sector to remove/treat the environmental burden. This approach was developed by Leontief himself, and he made use of it to carry out empirical analysis on air pollution (Leontief 1970). Moreover, Nakamura and Kondo (2002) produced an input-output table for waste based on this approach.

Here, the structure of an input-output model for environmental analysis is explained. It is based on the aforementioned first approach, which has a simpler model structure and a larger number of applied case studies.

In Formula 17.4, d_j stands for CO_2 emission directly emitted accompanying production activity from each industrial sector. Next, the emission coefficient (c_j), indicating emission per one unit of production, is derived. This is the total emission (d_j) in each industry divided by the input (x_j), and this process is shown in the following formula:

$$c_j = \frac{d_j}{x_j} \tag{17.4}$$

The total emission (d_j) in each industry can be described as the product of emission coefficient (c_j) and output (x_i), when the emission coefficient of Formula 17.4 is used. Based on the input-output table in Table 17.1, the CO_2 emission in each industry is shown below as an input-output model formula:

$$\begin{pmatrix} d_1 \\ d_2 \end{pmatrix} = \begin{pmatrix} c_1 & 0 \\ 0 & c_2 \end{pmatrix} \begin{pmatrix} x_1 \\ x_2 \end{pmatrix} \Leftrightarrow \mathbf{d_p} = \mathbf{Cx} \\ = \mathbf{C}\{\mathbf{I} - (\mathbf{I} - \mathbf{M})\mathbf{A}\}^{-1}\{(\mathbf{I} - \mathbf{M})\mathbf{f} + \mathbf{e}\} \tag{17.5}$$

where $\mathbf{d_p}$ in Formula 17.5 indicates direct CO_2 emission accompanying production in each industry and can be called production-based emission or direct emission. Of course **f** and **e** in Formula 17.5 are exogenous values, and measuring production-based emission depending on random production of **f** and **e** can be done.

On the other hand, deriving consumption-based emission ($\mathbf{d_c}$) can be done by a similar idea:

$$\underbrace{\mathbf{d_c} = \mathbf{C}\{\mathbf{I} - (\mathbf{I} - \mathbf{M})\mathbf{A}\}^{-1}(\mathbf{I} - \mathbf{M})\mathbf{f}}_{\text{Induced emission}} \quad \underbrace{+d_h}_{\text{Direct emission in household}} \tag{17.6}$$

The first item on the right side of Formula 17.6 is emission, including indirect emission, induced by production of final demand such as private consumption, and can be called induced emission. The $\{\mathbf{I} - (\mathbf{I} - \mathbf{M})\mathbf{A}\}^{-1}$ type of Leontief inverse is used in Formula 17.6, based on the import endogenous model; therefore, emission induced overseas based on import demand is not included. This induced emission

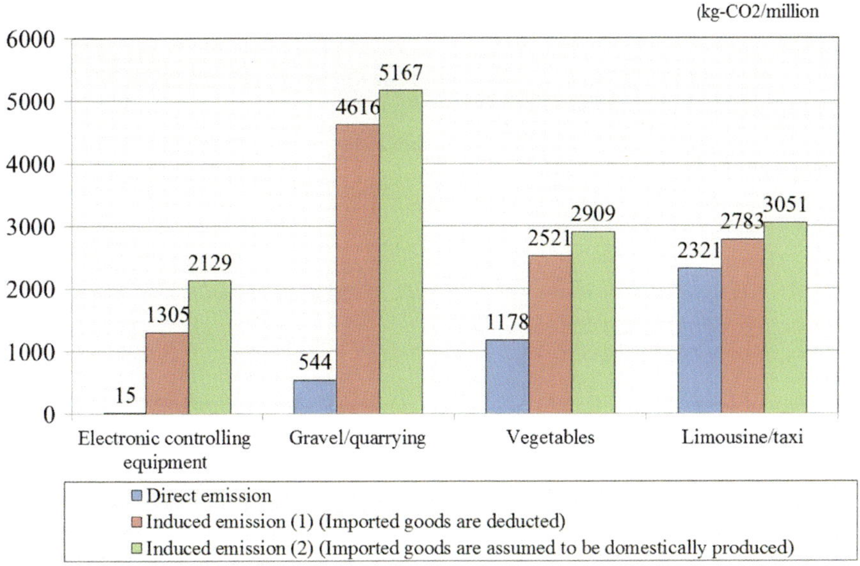

Fig. 17.2 CO_2 emission per production unit in industrial sectors in Japan (Source: Nansai and Moriguchi 2012)

with the addition of direct emission from households d_h can be considered consumption-based emission.

Next, direct emission (emission coefficient) and induced emission per unit of accompanying production, estimated in actual industrial sectors based on this idea, are compared and examined. Figure 17.2 shows direct CO_2 emission and induced CO_2 emission that is inevitable per unit (1 million yen) of production, in some industrial sectors as examples, using the aforementioned database of 2005 by the National Institute for Environmental Studies (Nansai and Moriguchi 2012).

Induced emission in Fig. 17.2 includes indirect emission, so we can see that it ends up being higher than direct emission in any industry. Moreover, induced emission (17.1) is measured by $\{\mathbf{I}-(\mathbf{I}-\mathbf{M})\mathbf{A}\}^{-1}$ type of Leontief inverse, and emission accompanying production of imported goods is deducted. On the other hand, induced emission (17.2) is calculated by $(\mathbf{I}-\mathbf{A})^{-1}$ type, including emissions due to imported goods, on the assumption that they are produced domestically. Note that induced emission (17.2) is higher than induced emission (17.1).

When direct emission is focused, it is lower in "gravel/quarrying" than it is in "vegetables" or "limousine/taxi." However, when compared with induced emission, that in "gravel/quarry" surpasses that of the other two industries, in fact, it is higher than in any industry in Fig. 17.2. When the ratio of direct emission and induced emission is focused, the balance between the two is relatively small in "limousine/taxi," while the balance is quite large in "electronic controlling equipment" where induced emission (17.2) is 138 times higher than direct emission. We can see that this ratio can vary widely among industries.

Examination of Fig. 17.2 suggests that there is a possibility of coming to the wrong conclusion when only direct emission of the industry is considered for evaluation of the burden that various industries place on the environment. Ripple effect analysis of environmental burden, based on the input-output method, indicates the importance of considering comprehensive impact on the environment including indirect impact.

17.1.3 Multiregional Input-Output Model and International Environmental Analysis

The type of input-output table explained so far is an input-output table that has the economy of a single country as its subject, but there are also input-output tables consisting of multiple countries or regions. Such input-output tables are called world input-output tables or interregional input-output tables, depending on the spatial levels. Generally, they are grouped under the blanket term of a multiregional input-output table (MRIO).

Suppose there is a world input-output table for country A and country B (Table 17.2) to explain the structure of the multiregional input-output table in Table 17.2, the transaction amount is divided into spatial level, and transactions between country A and country B are explained per sector. In such a multiregional input-output table, the transaction relation of economic flow is entered at the spatial level between two countries in addition to sectors of industry, household, and government. This enables clear understanding of which industry in which country

Table 17.2 Structure of a multiregional input-output table

		Intermediate demand		Final demand		Export to R.O.W.		Import from R.O.W		Production value
		Country A	Country B	Country A	Country B	Country A	Country B	Country A	Country B	
Intermediate input	Country A	Self-sufficiency of intermediate goods in Country A	Export of intermediate goods from Country A to Country B	Self-sufficiency of final goods in Country A	Export of final goods from Country A to Country B	Export of Country A		Import of Country A		Production value of Country A
	Country B	Export of intermediate goods from Country B to Country A	Self-sufficiency of intermediate goods in Country B	Export of final goods from Country B to Country A	Self-sufficiency of final goods in Country B		Export of Country B		Import of Country B	Production value of Country B
Added value		Added value of Country A	Added value of Country B							
Production value		Production value of Country A	Production value of Country B							

Notes: ROW stands for rest of world, indicating countries besides country A or country B. For example, "export of intermediate goods from country A to country B" can be phrased as "import from country A to country B." Therefore, the transaction amount can be read from the viewpoint of the imported amount in both countries

the intermediate goods and final goods were produced and, thus, in which of the two countries it should be input. This means imports and exports between the two countries become endogenous in the structure of the input-output table.

The multiregional input-output model can be applied to development of the input-output model and environmental analysis, as already mentioned. In particular, the necessity for investigating the impact of, and measures to deal with, environmental problems from a global viewpoint is becoming more important with the background of intensifying overseas expansion of companies and changes in the international division of labor. Therefore, international environmental analysis using the multiregional input-output model plays a very important role.

Analysis cases of CO_2 emission using the multiregional input-output model are explained below. Table 17.1 shows CO_2 emission of various countries, mainly in Asia, estimated by Shimoda et al. (2011) This study uses the Asian International Input-output Table 2000 made by the Institute of Developing Economies, Japan External Trade Organization. This table is a multiregional input-output table consisting of 76 industrial sectors in ten countries and regions of Indonesia, Malaysia, the Philippines, Singapore, Thailand, China, Taiwan, Korea, Japan, and the USA. It is widely used in studies concerning environmental and economic analysis, especially in Asia (e.g., Shimoda et al. 2011; Su and Ang 2011). The Institute of Developing Economies also makes the BRICs International Input-Output Table 2005 edition.

In Table 17.3, emission based on the demand of each country is entered vertically per supplier nation, and the vertical sum is, simultaneously, the total consumption-based emission and the sum of induced emission. Emission based on supply in each country is entered horizontally per consumer nation, and the horizontal sum is simultaneously the total production-based emission and direct emission emitted in each country. That is to say, Table 17.3 can be considered a

Table 17.3 Trade structure of CO_2 emission in Asian countries in 2000 (million tons-CO_2)

Emission based on demand →

Emission based on supply ↓	Indonesia	Malaysia	Philippines	Singapore	Thailand	China	Taiwan	Korea	Japan	U.S.A.	Others	Total (production base)
Indonesia	184	2	1	2	2	5	3	4	16	19	53	292
Malaysia	1	47	1	5	2	5	2	2	11	23	48	146
Philippines	0	0	49	0	0	1	1	1	3	8	13	76
Singapore	1	2	1	20	1	3	1	1	2	5	35	71
Thailand	1	1	1	1	104	3	1	1	8	13	42	174
China	5	5	2	5	6	2,466	10	20	104	199	399	3,221
Taiwan	1	1	1	1	1	14	143	2	8	20	56	248
Korea	2	1	1	1	1	20	4	356	18	29	103	537
Japan	1	2	1	2	2	8	5	5	919	34	70	1,051
U.S.A.	2	3	2	3	3	13	10	12	43	5,448	519	6,058
Total (consumption base)	197	65	59	40	123	2,537	180	403	1,134	5,797	1,339	11,873

Source: Table 8 (p. 215) in Shimoda et al. (2011)

matrix table indicating the trade balance of CO_2 emission by distributing the total emission of the ten subject countries, from the viewpoint of supply and demand.

For example, focus on Indonesia. Of the total emission of 292 million tons (sum of production base), directly emitted in Indonesia, emission produced in satisfying the demand from its own population is 184 million tons. Some 2 million tons were emission for demand in Malaysia and 1 million tons for demand in the Philippines; thus, we can interpret this to mean that 108 million tons of CO_2 were emitted in Indonesia to satisfy overseas demand. Hence, in the case of Indonesia, 37 % of direct emission was induced by overseas demand. This figure varies widely, from 72 % in Singapore to 10 % in the USA.

When the sums of both consumption base and production base are compared, the consumption base surpasses the production base only in Japan, making it the sole net importer of CO_2 emission among the ten countries in Table 17.3. However, Table 17.3 does not include consumption base emission based on imports from countries other than the included ten. Therefore, care should be taken: whether these countries are actually net importers or net exporters cannot be judged from Table 17.3 alone.

The trade structure of CO_2 emission mainly in Asia is examined based on the environmental analysis method, using the input-output model. The environmental burden generated by economic activities in one country is not limited to that country alone, but has impacts on various countries. This phenomenon is sometimes called carbon leakage in the case of CO_2 emission. Carbon leakage among Asian countries is tending to increase due to recent rapid economic development and more advanced and complicated international division of labor. A phenomenon such as carbon leakage is one of the factors that obscure the causes and structure of environmental problems in Asia and that make solution of environmental problems more problematic. In such circumstances, further development and application of environmental analysis methodology, using the multiregional input-output model, is required to implement effective environmental analysis and policy recommendations.

References

Leontief W (1970) Environmental repercussions and the economic structure: an input-output approach. Rev Econ Stat 52(3):262–271. http://www.cger.nies.go.jp/publications/report/d031. Viewed 19 Mar 2014

Miller RE, Blair PD (2009) Input-output analysis: foundation and extensions second edition. Prentice-Hall, Englewood Cliffs

Nakamura S, Kondo Y (2002) Input-output analysis of waste management. J Ind Ecol 6:39–64

Nansai K, Moriguchi Y (2012) Embodied energy and emission intensity data for Japan using input-output tables (3EID) – web edition. National Institute for Environmental Studies, Tsukuba, Japan

Shimoda, M, Watanabe T, Ye Z, Fujikawa K (2011) An empirical study on interdependency of environmental load and international I-O structure in the Asia-Pacific Region. In: Ji H, Onishi

H (eds) Japan China economic statistics review. Capital University of Economics and Business Press, Beijing, China, pp 200–220
Shishido S (editorial supervision) Pan Pacific Association of Input-output Studies (ed) (2010) Input-output analysis handbook. Toyo Keizai Inc. (in Japanese)
Su B, Ang BW (2011) Multi-region input–output analysis of CO_2 emissions embodied in trade: the feedback effects. Ecol Econ 71:42–53

Chapter 18
Corporate Environmental Management and Environmental Strategies

Hidemichi Fujii

Abstract This chapter focuses on environmental management, corporations' approaches to pollutant emission reduction, and the introduction of incentives for corporations to reduce pollutant emissions. It also discusses the effectiveness of models of these approaches, the burden of expense, and the efficiency of these approaches. In particular, the emphasis is on pollution abatement measures in the production process, with detailed explanations of the features and differences between end-of-pipe type emission reduction measurements and cleaner production type emission reduction measurements. In addition, factor analysis models that use numerical data to calculate the amount of (and changes in) pollutant emissions are introduced.

Keywords Environmental management • Factor analysis • Manufacturing sector • Toxic chemicals • Cleaner production • End of pipe

18.1 Environmental Management Toward a Sustainable Society

Rapid technological progress and economic development has enabled people to achieve a convenient and affluent lifestyle. At the same time, we have imposed a burden on the global environment through this lifestyle, often expressed in terms of mass production, mass consumption, and mass disposal. As a result, various environmental problems have intensified including air pollution, water pollution, global warming, desertification, and a decrease in biodiversity. Without taking certain measures, it will be difficult to achieve sustainable development, and we will leave the next generation with a heavy burden to bear.

On the other hand, movement toward environmental preservation is increasing globally. In 1987, the World Commission on Environment and Development (Brundtland Commission) report presented sustainable development as the

H. Fujii (✉)
Graduate School of Fisheries and Environmental Sciences, Nagasaki University, Nagasaki, Japan
e-mail: hidemichifujii@gmail.com

T. Shimaoka et al. (eds.), *Basic Studies in Environmental Knowledge, Technology, Evaluation, and Strategy*, DOI 10.1007/978-4-431-55819-4_18

fundamental policy of development. The Kyoto Conference to stop global warming was held 10 years later (in 1997), and the Kyoto Protocol, which set a prescribed rate of greenhouse gas reduction, was signed. The Kyoto Protocol came into effect when ratified by Russia on February 16, 2005, making reduction targets obligatory for developed countries in the First Commitment Period from 2008 to 2012.

In this context, corporate activity plays a role in the decision-making that strikes a satisfactory balance between the environment and the economy and which is crucial in achieving sustainable development in society. This is apparent because the economics of development and the origin of economic value are corporate activities, while at the same time, a great deal of the environmental burden is caused by this corporate activity (Fujii et al. 2013b). The aim of corporate management is the pursuit of profit, and until relatively recently, it was considered that environmental measures did not concur with corporate management. Along with the increased attention on environmental problems in society, the evaluation axis of environmental preservation was added to the needs of the market and its customers, and the demand for corporations to take on environmental measures has intensified. In response to such demand, environmental management is attracting attention as a model for corporate management that takes environmental problems into consideration (Fujii and Kimbara 2013).

Environmental management is a strategy aimed at securing a balance between environmental preservation and business management. A considerable number of corporations today acknowledge that environmental management is an essential strategy for sustainable corporate management. In this chapter, the focus is on the environmental management of corporations, and attention is devoted to antipollution measures and their cost.

18.2 Measures Against Pollution in the Manufacturing Process

Corporate economic activity is inevitably accompanied by direct and indirect environmental burdens. Figure 18.1 shows the relationship between raw material consumption and environmental pollution. The term "direct environmental burden" refers to the amount of emitted pollutants generated in the manufacturing process, while the term "indirect environmental burden" refers to pollutants contained within the raw materials used.

Environmental pollution emitted by corporations is determined by the amount of pollution generated during the consumption of raw materials and the retrieved/removed amount of generated pollutants. The amount of environmental pollution produced by a corporation depends on the volume of raw materials consumed. The more pollutant-containing raw materials inputted, the more environmental pollution generated; conversely, input of fewer polluting raw materials leads to the generation of less environmental pollution. Therefore, it is important to economize

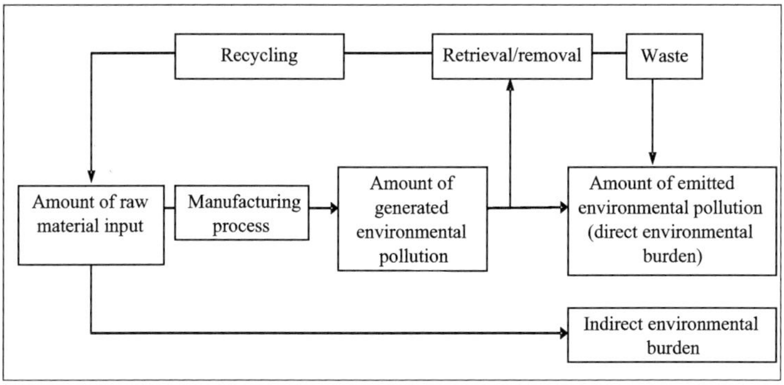

Fig. 18.1 Relationship between raw material consumption and environmental burden (*Source*: Made by the author)

on the raw materials used, while generated pollutants should be retrieved or removed as a way to reduce the environmental burden accompanying the economic activity of production. The former approach achieves economies in the amount of raw materials consumed by making the manufacturing process more efficient via energy or resource conservation, while the retrieval/removal process of pollutants is divided into measures to transform collected waste into materials (i.e., recycling) and discharging waste outside the corporation without any recycling. A case of using recycled industrial wastewater generated through water resource consumption as circulating water is included in the former, while the latter includes waste (sludge) generated as a result of the retrieval/removal of sulfur dioxide by flue gas desulfurization equipment.

Here, measures against emissions of chemical substances are explained. Many newly industrialized countries experienced pollution problems, including water pollution and air pollution, during the transition period from early economic development to industrialization. End-of-pipe (EOP) type measures were taken to combat these environmental problems (Cagno et al. 2005). EOP type measures are environmental measures involving direct symptomatic treatment, such as the removal of toxic chemicals by installing filters at the outlet of smokestacks and drainpipes. Typical measures are seen in flue gas desulfurization equipment and wastewater treatment equipment. Drawbacks to the EOP type measures include the possibility of generating secondary pollution (e.g., sludge waste generated in wastewater treatment) and the costs of installation and operation.

In contrast to these EOP measures, there are cleaner production (CP) type measures where the amount of raw materials put in is economized by optimizing the manufacturing process, thus leading to a reduction in toxic chemicals. Typical CP measures are seen in ecodesigning when the product design is made more efficient by economizing the amount of raw materials used or where less environmentally stressful materials are substituted for those traditionally used in the production process. CP type measures, similar to EOP type measures, require a certain amount of initial investment, but the operational cost is lower because they do not require filters or absorbents to remove pollutants. Moreover, simplifying and optimizing the production process can save raw material and labor costs, and as a result, reduced manufacturing costs can be expected (Baas 1995).

It is impossible, however, to reduce all pollutants by CP type measures, and there is a limit to the types of pollutants that can be reduced. EOP type measures often need to be combined with CP type measures to satisfy the environmental standard or a voluntary standard established by corporations (Frondel et al. 2007). Therefore, the balance between the two types of measures is important. The promotion of CP is indicated in Agenda 21, which was adopted at the Earth Summit in 1992, and today, development and distribution of technical information is being worked on in numerous countries through various projects.

18.3 Measures Against Pollution and Cost Allocation

As previously mentioned, there are numerous types of approaches for environmental pollution abatement measures, and their costs and effects vary among industries. Table 18.1 shows industrial data related to pollution abatement cost and expenditures (PACE) among US corporations in FY 2005; these data were presented by the US Bureau of the Census in 2008. The table shows shipment value, total cost of measures, and the rate of the total cost of measures divided by the shipment value in each industry.

The percentage rates in Table 18.1 show that the cost allocation ratio is higher in basic material industries such as textile mills and paper, chemical, nonmetallic mineral product, and primary metal manufacturing, while fabricating and assembly industries such as machinery, computer and electronic products, and transportation equipment manufacturing have lower cost allocation ratios. In addition, the cost breakdown (activity) indicates that basic material industries have higher ratios in "treatment cost," signifying EOP type measures, while fabricating and assembly industries show a higher cost allocation for "recycling cost" and "disposal cost."

As seen here, the specific nature of each industry makes a considerable difference concerning applicable measures against pollution and the cost of environmental pollution abatement measures. This is because the types and quantities of chemicals used in manufacturing processes differ among industries, and this fact has a major impact on the ease of recycling or difficulty in disposal.

If the government were to introduce a policy to promote CP type measures, how would the effect of such an introduction be verified? Conducting individual inspections of pollution measures at each corporation requires a considerable amount of time and energy and is therefore difficult. Under such circumstances, an analysis model can be introduced that distinguishes whether a certain pollution measure is being implemented in a CP type or an EOP type measure by using pollution emission data.

Table 18.1 Comparison of pollution abatement cost and expenditure by type of industry in the USA, FY 2005

	(1) Annual Survey of Manufactures (ASM) value of shipments (million USD)	(2) Total cost (million USD)	Ratio	Breakdown of cost and expenditure by activity			
			[(2)/(1)] (%)	Treatment (%)	Prevention (%)	Recycling (%)	Disposal (%)
All industries	4,735,384	20,678	0.44	52	17	8	22
Food	534,878	1,573	0.29	55	11	7	28
Textile mills	41,149	221	0.54	63	7	9	21
Paper	162,848	1,796	1.10	60	11	7	23
Petroleum and coal	476,075	3,746	0.79	51	35	7	8
Chemical	604,501	5,217	0.86	53	16	8	24
Plastics and rubber	200,489	503	0.25	43	16	10	32
Nonmetallic mineral	114,321	696	0.61	57	18	7	18
Primary metal	201,836	2,291	1.14	54	12	10	24
Machinery	302,204	316	0.10	34	16	11	39
Electronic product	373,932	624	0.17	54	9	10	27
Transportation equipment	687,288	1,319	0.19	45	13	12	30

Source: United States Bureau of the Census 2008 "Pollution abatement costs and expenditures: 2005"

Table 18.2 Sample of the data set for analysis

	(A) Amount of production	(B) Amount of pollutant emission	(C) Amount of retrieved pollutants	(D) Generated pollutant amount [(B) + (C)]
2013	100	100	100	200
2014	200	120	200	320

18.4 Factor Analysis of Environmental Pollutant Emission

The preceding chapters posited that there are EOP type measures and CP type measures to address pollution. In this chapter, an analysis method to determine the suppression effect on pollutant emission involved in each approach is introduced. The data set shown in Table 18.2 is used to explain the analysis method.

Table 18.2 shows data related to the (A) amount of manufactured products, (B) amount of emitted pollutants, and (C) amount of retrieved pollutants from 2013 to 2014. Here, the sum of (B) amount of pollutant emission and (C) amount of retrieved pollutants can be considered as the amount of pollutants generated in the production process. Therefore, (D), the generated pollutant amount, is entered in the table.

Next, two indices that are necessary for the analysis are explained. The first is an index concerning CP type measures. Because CP type measures are those that suppress the generation of pollutants during the product manufacturing process, a comparison of the number of products produced and the amount of pollutants generated can be used to assess the effect of these measures. Therefore, the index to verify CP type measures is as follows: (D) generated pollutant amount/(A) amount of production. A decrease in this index indicates the level of achievement in reducing pollutant generation while still maintaining production.

The second is an index to assess EOP type measures. Because the EOP type is used to measure the retrieval of generated pollutants, assessment can be performed by comparing the generated amount and the retrieved amount of pollutants. Therefore, the index to verify EOP type measures is as follows: (B) amount of pollutant emission/(D) generated pollutant amount. A decrease in this index indicates the level of achievement in reducing pollutant emission by increasing the retrieved amount without reducing the generated amount of pollutants. By using the two indices mentioned above, pollution emission can be indicated by the following formula (18.1):

$$(\mathrm{B}) = (\mathrm{B})/(\mathrm{D}) \times (\mathrm{D})/(\mathrm{A}) \times (\mathrm{A}) \tag{18.1}$$

This formula indicates that pollution emission can be analyzed by the three factors: an index to assess CP type measures (generated pollutant amount/amount of production), an index to assess EOP type measures (amount of pollutant emission/amount of production), and amount of production. Amount of production is used as an index to show the scale of production. This is because the generated

Table 18.3 Factor analysis of the amount of emission

	(B) Amount of pollutant emission	(B) Amount of pollutant emission/(D) generated pollutant amount	(D) Generated pollutant amount/(A) amount of production	(A) Amount of production
2013	100	0.500	2.00	100
2014	120	0.375	1.60	200
Ratio	120 %	75 %	80 %	200 %

amount of pollutants is affected by the production scale, which depends on the amount of pollutant input from the raw materials.

The result of the calculation including the above indices is entered in Table 18.3. Here, the ratio is computed by dividing the numbers from 2014 by those from 2013 to examine the change in numbers from 2013 to 2014.

Table 18.3 indicates that pollution emission increased by a factor of 1.2 from 2013 to 2014. Next, the individual factors are examined to establish the cause of the emission increase. The index of (B) amount of pollutant emission/(D) generated pollutant amount showing EOP type measures has a ratio of 75 %. Therefore, a reduction down to 75 % of emission per generated amount was achieved year to year (= 25 % reduction) by increasing the retrieved amount of pollutants.

In addition, the index of (D) generated pollutant amount/(A) amount of production showing CP type measures has a ratio of 80 %. Therefore, a reduction down to 80 % pollution generation per unit production was achieved year to year (=20 % reduction) by utilizing production processes and product designs that do not generate as much pollution.

From these statistics, it is apparent that EOP type measures and CP type measures are effective in reducing pollution emission. Meanwhile, (A) amount of production, which illustrates the level of production, doubled from 1 year to the next, and it is clear that the expansion of production was a major factor in the increase of emissions.

As we have seen, it is possible to identify how the change in emission was affected by the factors of CP type measures, EOP type measures, and production size by applying the factor analysis model in the emission of pollutants. Applied models concerning factor analysis are detailed in Fujii and Managi (2013) and Fujii et al. (2013a).

18.5 Summary

This chapter focuses on environmental management and approaches to pollutant emission reduction by corporations. It also discusses the effectiveness of models regarding approaches, the burden of expense, and the efficiency involved in such approaches. Corporate measures to combat pollution are becoming increasingly essential, and demands by shareholders and consumers are gaining momentum. On the other hand, any cost allocated for pollution measures does not directly contribute to increased production or improved profitability, and it can be difficult for corporations to adopt a management strategy that promotes assertive measures to address pollution. At the same time, CP type measures are measures that reduce pollution while environmental preservation and economic efficiency are simultaneously achieved, providing an incentive to corporations for capital investment and input of human resources. Ultimately, for corporations to promote pollution abatement measures, it is important for them to understand how CP type measures can be implemented.

By applying the factor analysis model of pollution emissions as a method, as it was introduced in this chapter, it becomes possible for corporations and business offices to identify the factors that have an impact on the change in pollutant emission. Accordingly, it is expected that the analysis results can aid in understanding the current condition of pollution abatement measures when examined by decision-makers in corporate offices. These results can be utilized as indices for strategy setting and goal setting for any pollution abatement measures taken in the future.

Today, consumers and shareholders in developed countries pay close attention to environmental preservation by corporations, and active engagement in dealing with environmental problems is likely to improve a corporate image. Meanwhile, in developing countries with a low average annual income and where securing foodstuffs and daily necessities is the foremost priority, corporate brand image or attitude toward environmental problems is hardly considered when making choices. Product pricing is the sole decision-making factor for competitiveness in those markets. For these markets and for consumers with a low level of awareness toward environmental issues, sincerely working on environmental preservation and disclosing the results are not likely to aid in gaining competitiveness. Therefore, there will likely be low motivation to actively take measures to benefit the environment if it involves additional cost and labor.

Consequently, it is important to provide incentives for environmental management to a degree that will increase profit when corporations in developing countries make the decision to introduce environmental management practices. Actively disclosing information on CP type measures is important for reaching such goals.

References

Baas LW (1995) Cleaner production: beyond projects. J Clean Prod 3(1–2):55–59

Cagno E, Trucco P, Tardini L (2005) Cleaner production and profitability: analysis of 134 industrial pollution prevention (P2) project reports. J Clean Prod 13:593–605

Frondel M, Horbach J, Rennings K (2007) End-of-pipe or cleaner production? An empirical comparison of environmental innovation decisions across OECD countries. Bus Strateg Environ 16:571–584

Fujii H, Kimbara T (2013) Environmental management and external factors of manufacturers in Japan and the U.S.A. Org Sci 46(4):83–101

Fujii H, Managi S (2013) Decomposition of toxic chemical substance management in three U.S. manufacturing sectors from 1991 to 2008. J Ind Ecol 17(3):461–471, Wiley-Blackwell

Fujii H, Managi S, Kaneko S (2013a) Decomposition analysis of air pollution abatement in China: empirical study for ten industrial sectors from 1998 to 2009. J Clean Prod 59(15):22–31, Elsevier

Fujii H, Iwata K, Kaneko S, Managi S (2013b) Corporate environmental and economic performances of Japanese manufacturing firms: empirical study for sustainable development. Bus Strateg Environ 22(3):187–201, Wiley-Blackwell

U.S. Environmental Protection Agency (2008) National Center for Environmental Economics, Pollution Abatement Costs and Expenditures: 2005 Survey. http://yosemite.epa.gov/ee/epa/eed.nsf/pages/pace2005.html. Accessed on 5 Mar 2014

Chapter 19
Basic of Project Management and Environmental Project Cases

Mami Shinozaki, Kenichi Tsukahara, Masaki Yokota, and Takeru Sakai

Abstract This chapter gives an outline of the basic knowledge of project management while introducing the main points of environmental project cases and case methods. Practical skills in project management methodology, which is management thinking, are required to execute environmental projects along with a deep knowledge of specific expertise. Moreover, it is important to have the ability to take a commanding view and grasp environmental issues in a wide and interdisciplinary manner. Furthermore, the requirements of an environmental project manager include strategic innovation ability in project design, consensus formation skills among a wide variety of stakeholders, and facilitation skills, among others. This chapter gives an explanation of basic vocabulary in project management in Sect. 19.1, the main points of environmental project cases in Sect. 19.2, and case methods in Sect. 19.3.

Keywords Environmental issues • Environmental project • Project management • PCM • *PMBOK Guide* • Strategic innovation ability for problem solution

19.1 Basic Vocabulary in Project Management

19.1.1 What Is a Project?

In *A Guide to the Project Management Body of Knowledge* (*PMBOK*[1] *Guide*), a project is defined as "a temporary endeavor undertaken to create a unique product,

[1] *A Guide to the Project Management Body of Knowledge*. This is a standard and guide recognized by specialists for project management regardless of the industry. It is published by the Project Management Institute (USA) in various languages in the world and used worldwide as well.

M. Shinozaki (✉)
Research Institute for East Asia Environments, Kyushu University, Fukuoka, Japan
e-mail: rieae.shinozaki@gmail.com

K. Tsukahara • M. Yokota
Faculty of Engineering, Kyushu University, Fukuoka, Japan

T. Sakai
New Campus Planning office, Kyushu University, Fukuoka, Japan

T. Shimaoka et al. (eds.), *Basic Studies in Environmental Knowledge, Technology, Evaluation, and Strategy*, DOI 10.1007/978-4-431-55819-4_19

service, or result." Interpretations of a project vary, and here a project is defined as "a unique and innovativeness planned over a definite term by human being, single or multiple organization(s) which sets up a goal to be achieved, based on a clear strategy.[2]" The project is clearly distinguished from routine work with continuity and repetition, thus positioned as irregular work. In addition, multiple stakeholders are involved with a project directly and indirectly, carrying out important tasks. Usually, the team members of a project are made up of experts from diverse fields required in the management, though this depends on the size of the project.

There are various types of projects: projects in the business management field including production project, new product development project, information system project, and feasibility study (F/S) projects; infrastructure projects such as construction project and urban development project; investment projects undertaken by a governmental bank or international agency; and research and development project held at research institutes and universities funded by the national government or international agencies. International projects include those held in a country other than their own and funded by governmental agency procurement, such as Official Development Assistance (ODA)[3] loan projects where domestic and international various agencies are involved.

In the background of initiating/planning a project, there are needs in the market/organization, needs of customers, technology development, response by legal regulation, and impact on the ecological system. In addition, there are the needs of the local government in the field of the project in the case of an international undertaking.

19.1.2 What Is a Project Management?

Project management is defined as "the application of knowledge, skills, tools and techniques to project activities to meet the project requirements" in the *PMBOK Guide*. Project management means "managing a project."

Then, what is management? In a management theory stemming from the study of business administration, there is a clear distinction between *kanri* which is a management in Japanese and management in English. Management in Japanese *kanri* has the strong connotation of "control." Management in English is not only to control but also to analyze, assess, improve, adjust, and strategically lead and control factors by way of decision making. The factors are various resources and assets, for example, "people," "products," "money," "information," and "natural

[2] Though there is an established distinction between project and program, they are often used interchangeably in Japan.

[3] Official Development Assistance means cooperation by providing money, technology, and human resources in order for the development of economy and society, improvement of social welfare, and measures for environmental problems.

environment." Those who manage projects can maintain all the factors in balance, lead the project to success, and take the responsibility in order to achieve a clear goal based on multiple risks, that is, project managers. Project managers are required to have leadership skills, consensus formation skills to negotiate with a diverse range of stakeholders, decision-making skills which translate as execution ability and humanity shown in facilitation skills, and strategic innovation skills in project design, let alone knowledge and experiences in projects. For team member, what brings about success with projects is having human resources and team member with ample expertise and experience in the required field to form a professional group with clear boundaries of expertise and responsibility.

The *PMBOK Guide* (Project Management Institute 2013) describes the nature of project management processes in terms of the integration between the processes, their interactions, and the purposes they serve. Project management processes are grouped into five categories known as "Project Management Process Groups (or Process Groups)": "Initiating Process Group," "Planning Process Group," "Executing Process Group," "Monitoring and Controlling Process Group," and "Closing Process Group." The 47 project management processes identified in the *PMBOK Guide* are further grouped into ten separate "Knowledge Areas": "Project Integration Management," "Project Scope Management," "Project Time Management," "Project Cost Management," "Project Quality Management," "Project Human Resource Management," "Project Communications Management," "Project Risk Management," "Project Procurement Management," and "Project Stakeholder Management," each of which is an essential project management area for the success of a project.

By identifying these ten Knowledge Areas at the planning phase, a project can be updated with minimum process if an unforeseeable accident happened in the executing process.

19.1.3 Method of Project Planning

It is necessary to clarify definition of factors and to draft the implementation plan by a project planning/proposal. The defining factor is composed of the following steps. WBS[4] is a useful method to identify work procedure.

1. Setting up the theme
2. Clarifying the guideline
3. Clarifying the course and project product/output
4. Analyzing the current situation at the project field
5. Making a project proposal

[4] Work breakdown structure: All the works necessary in project activity are divided into manageable units for breakdown in order to clarify the work items.

Project Cycle Management (PCM)[5] method has been widely used by the Japan International Cooperation Agency (JICA) as one which was available among major environmental projects for development assistance in Asian countries. The PCM method is used in problem-solving projects subjected to development assistance projects involving international technology cooperation using Japanese yen-denominated government credit "Japanese Official Development Assistance (ODA)."

The PCM method was developed on the basis of a concept which interprets a project as the entire cycles, which are termed "Planning", "Implementation" and "Evaluation". This method consists of "Participatory Planning" to work out a solution and "Monitoring and Evaluation". Here, we are focused on "Participatory Planning," which is a plan making for a solution. To map out a solution with involved local government officials by the PCM method, it becomes feasible to organize a solution that is not inclined to specific needs and for associated stakeholders to share collective recognition by "Stakeholder Analysis", "Problem Analysis" and "Objective Analysis." Project Design Matrix (PDM) is a standardized format of the project management that has shown in a general way the operating and managing of the project after three analyses are complete.

Characteristics and Advantages of the PCM method

In "PCM-Management Tool for Development Assistance (Participatory Planning), 7th edition", Foundation for Advanced Studies on International Development (FASID) (2008), the *characteristics of the PCM method* are defined as follows:

1. Participatory Approach
2. Logicality
3. Consistency

It takes a "participation" workshop style involving the stakeholders. A project can be organized by the local concerned people for discussion and ensures equal respect for the opinions of different groups. When the stakeholder is subjectively involved in a project from start to end by PCM, they can find the accurate problems and feasible solutions in the analytical process logicality. Also, they can be executing the project by themselves using the PDM which is consisted of planning, implementation, and evaluation.

The *advantages* of the PCM method are as follows:

1. Accurate and effective project management
2. Project planning that reflected recipients' needs
3. Improved transparency of development assistance
4. Effective communication among stakeholders

---

[5] Project Cycle Management: It was developed by the Foundation for Advanced Studies on International Development in reference to goal-oriented project planning method of the German Technical Cooperation.

Flow of the PCM method

1. Stakeholder Analysis: method of digging up and analyzing stakeholders
2. Problem Analysis: method of analyzing the problems by applying "Causes and Effects" relation
3. Objective Analysis: method of analyzing solutions by applying "Means and Ends" relation
4. Project Selection: confirming priorities in projects and selecting one project
5. Formulation of PDM: made by formulating a plan to acquire a budget

19.2 Case Study of Environmental Project

In environmental project management education in Kyushu University, it is outlined that environmental project cases are used as the teaching materials. The purpose of this section is to aim for strategic innovation ability of problem solutions at the project planning stage and to understand how risk management is applied in project.

19.2.1 Case Study of Water Supply System Management Project: Comparative Study on Water Supply System Management Focusing on a Nonrevenue Water Ratio in Post-privatization Metropolitan Manila and Post-1950's Fukuoka City (Tsukahara et al. 2008)

Efficient water supply system management is a key to achieve the tenth goal of the Millennium Development Goals to reduce by half the proportion of population of people without sustainable access to safe drinking water. The objective of this project is to identify potential solutions to achieve sustainable water supply system management through comparative analysis on water supply management systems of post-privatization Metropolitan Manila (1994–2005) and post-World War Fukuoka City (1950's) is focusing on reducing illegal connections and water leakage. The result shows that the two cases are comparable in their revenue water ratios, investment levels, tariff levels, and user's income levels, and substantive investment in water mains to reduce illegal connections and water leakage is a key to achieve sustainable water supply system management. The result also shows that sustainable water supply system management can be achieved about 70 % revenue water ratio both in two cases and the 70 % revenue water ratio is a feasible target for developing countries where their income levels are about US 1,000 dollar.

Objective of the Project

To show the importance of improvement of a revenue water ratio (RWR) is to realize sustainable water supply system management in developing countries. The objective of this project is to identify potential solutions to achieve sustainable water supply system management through comparative analysis on water supply management systems of post-privatization Metropolitan Manila (1994–2005) and post-World War Fukuoka City (1950's) is focusing on reducing illegal connections and water leakage.

Issues of the Project

Manila Water Company Inc. (MWCI), Eastern Metropolitan Manila (MM), succeeded to improve a business and its RWR to 70 % from above 40 % by reforming the pipe connections because of drastic water leakage and water theft prevention measure since 2002. On the other hand, Maynilad Water Services Inc. (MWSI), Western MM, was not successful in improving its RWR and went out of business in 2006.

Team of the Project

Kyushu University, Fukuoka Asian Urban Research Center (URC), Japan International Cooperation Agency (JICA), Metropolitan Waterworks and Sewerage System (MWSS), Manila Water Company Inc. (MWCI), Maynilad Water Services Inc. (MWSI), and Nippon Jogesuido Sekkei Co., Ltd. (NJS)

Contents of the Project

Situations in Metropolitan Manila Water Supply System

In MM, water supply system development started in the late nineteenth century. In 1995, under the management of the Metropolitan Waterworks and Sewerage System (MWSS), its service coverage was 60 % in population. At the same time, its RWR was quite at a low level of 39 %. Due to inefficient management and its huge amount of accumulated debt, MM water supply and sanitation system was divided to two parts (east and west) and privatized in 1997. These two privatized companies are MWCI and MWSI and are called concessionaires.

Water sources of MM are mainly the Angat River and the Ipo Dam. As of 2007, water supply capacity and service area of MWCI and MWSI are 1,600 million liters per day (MLD) and 540 km^2 and 2,400 MLD and 1,401 km2, respectively. The boundary between MWCI and MWSI service area is shown in Fig. 19.1. Before

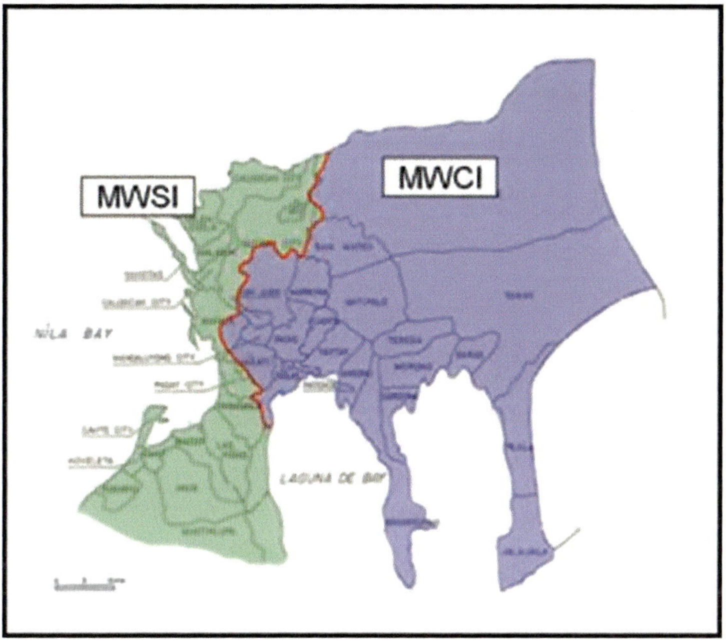

Fig. 19.1 Map of Metropolitan Manila and areas of MWCI and MWSI

privatization of MWSS, several official development assistance projects were launched. Angat Water Supply Optimization Project and Umiray-Angat Transbasin Project were the biggest projects in terms of investment scale. Total cost of these two projects was about 450 million US dollar (USD). In conjunction with other projects, these two projects achieved 4,000 MLD water supply capacity in MM.

Features of Fukuoka City Water Supply System Management in the 1950's

Fukuoka City water supply system was founded in 1923 as a public utility managed by the Fukuoka City government. A feature of Fukuoka City is its scarcity of water sources; thus, from the beginning of its operation, water source development and efficient use of water sources have been important.

Overview of Fukuoka City water supply system from 1950 to 1959 is shown in Table 19.1. Fukuoka City water supply system was severely damaged by the air attacks in 1945; leakage ratio of the water supply system became about 60 % at that time. From 1945 to 1948, emergent rehabilitation works were conducted to the water supply system. In addition, the first five-year anti-leakage project was conducted from 1956 to 1960. Due to these measures, the RWR of Fukuoka City water supply system improved from 45.7 % (1950) to 70.8 % (1959).

Table 19.1 Overview of Fukuoka City water supply system in the 1950's

Year	Served population	Served population Ratio (%)	Per capita average Volume of supply (litter)	Revenue Water ratio (%)
1950	304,705	77.6	169	45.7
1951	321,456	79.7	169	51.6
1952	329,297	78.0	169	53.0
1953	343,941	78.0	164	64.6
1954	352,909	72.7	178	61.6
1955	367,726	66.8	193	61.3
1956	385,097	66.5	199	61.2
1957	421,836	69.9	186	65.2
1958	438,583	71.5	181	67.6
1959	463,953	73.2	181	70.8

Table 19.2 Features of water supply systems and economic situations of Fukuoka City (1950's) and MM

	Fukuoka City	Metro Manila
Per capita income (USD purchasing power parity in 1990)	1,873 (as of 1950)	2,213 (as of 1992)
RWR	45 % (actual 1950)	40 % (actual 1995 MWSS)
	70 % (actual 1959)	65 % (actual 2005 MWCI)
		70 % (target in 2015)
Served population ratio (%)	60 % (actual 1950)	60 % (actual 1995 MWSS)
	70 % (actual 1960)	70 % (target in 2015)
Water tariff for 30 m^3 per month	430 yen (1959)	326 pesos (MWCI 2005)
Average monthly income	15,000 yen (1959)	9,950 pesos (2006)
Share of water tariff in monthly income	2.9 %	3.3 %

Comparison Between 1950's Fukuoka City Water Supply System and That of Post-privatization Metropolitan Manila

Table 19.2 shows features of water supply systems and economic situations of Fukuoka City (1950's) and MM water supply system (1997–2005). Major features are as follows:

RWR: RWRs of Fukuoka City in 1950 and 1959 were 45 % and 70 %, respectively. RWR of MM in 1995, RWR of MWCI in 2005, and target RWR of MM in 2015 were 40 %, 65 %, and 70 %, respectively. This shows that Fukuoka City (1950's) and MWCI (1995–2005) achieved similar level of improvement of RWRs.

Service coverage ratio: Service coverage ratios of Fukuoka City (1950) and MM (1995) were equally 60 %. This shows that there was not much difference between Fukuoka City and MM water service systems in view of the service coverage ratio.

Income level: Per capita income levels of Japan in 1950 and that of the Philippines in 1992 were 1,873 and 2,213 USD, respectively, in terms of purchasing power parity rate of 1990. This shows that these two situations are comparable in view of income level.

Tariff level: Tariffs of 30 m^3 (household use) in Fukuoka City (1959) and MWCI (2005) were 430 yen and 326 pesos, respectively. Other statistics show that ordinary workers' monthly payments in Fukuoka City (1959) and MM (2006) were 15,000 yen and 9,950 pesos, respectively. From these, the shares of water tariff in ordinary workers' income of Fukuoka City (1959) and MM (2006) were 2.9 % and 3.3 %, respectively. This also shows that these two situations are comparable in view of tariff level in household income.

Water sources: Both in Fukuoka City and MM, there are no big rivers nearby. Therefore, both Fukuoka City and MM needed to develop new water sources such as dams and transbasin projects. Fukuoka City and MM are comparable in this regard.

Figure 19.2 shows the 9-year changes in RWRs of Fukuoka City, MWCI, and MWSI, respectively. This shows that MWSI's management failed for its RWRs remain at low levels of below 40 %, while MWCI and Fukuoka City reach about 70 % RWRs. The fact that MWCI and 1950's Fukuoka City achieved sustainable management of water supply system supports the mainstream vision of "RWR improvement is a key to sustainable management of a water supply system."

Figure 19.3 shows that the yearly capital expenditures of Fukuoka City (late 1950's) and those of MWCI (2004 and 2005) were almost the same level, while those of MWSI were substantively low level.

Many staffs of water services in developing countries mention that "Rich countries such like Japan can achieve high RWRs due to their economic capacity, while "poor" countries cannot afford such investment." This section showed that Fukuoka City (1950's) and MWCI achieved improvement of RWRs by appropriate

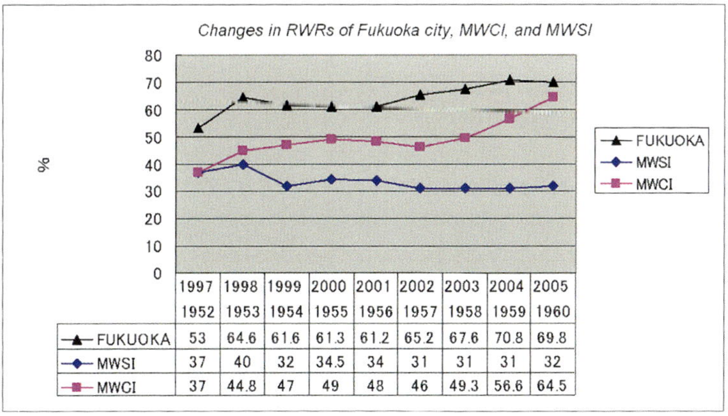

	1997	1998	1999	2000	2001	2002	2003	2004	2005
	1952	1953	1954	1955	1956	1957	1958	1959	1960
FUKUOKA	53	64.6	61.6	61.3	61.2	65.2	67.6	70.8	69.8
MWSI	37	40	32	34.5	34	31	31	31	32
MWCI	37	44.8	47	49	48	46	49.3	56.6	64.5

Fig. 19.2 Changes in RWRs of Fukuoka City, MWCI, and MWSI

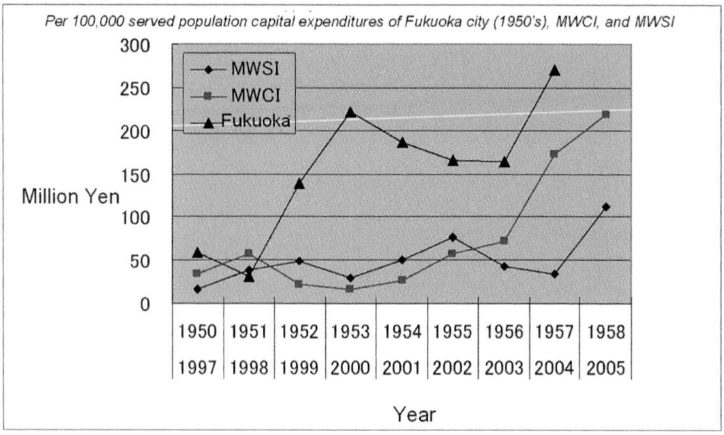

Fig. 19.3 Per 100,000 served population capital expenditures of Fukuoka City (1950's), MWCI, and MWSI

investments for improving RWRs even though their income levels were about 2,000 USD in terms of per capita purchasing power parity of 1990. This implies that achieving high RWR and sustainable management of water supply system is not an unrealistic target for developing countries.

Method and Outputs of the Project

It shows three examples of water supply system development and management: successful one by privatized company in a developing country, unsuccessful one by privatized company in a developing country, and successful one by publicly operated utility, where their income levels are about 2,000 USD in terms of purchasing power parity in 1990. The result shows that it is feasible for water supply system entities to supply water with affordable price if they manage water supply system properly. Major findings of this study are described as follows:

1. There is a view that Japanese water supply system management is too advanced to apply to developing countries. However, this study shows that Fukuoka City's water supply system management experience in 1950's can be applicable to current developing countries as its economy level was at the time similar to those of current developing countries.
2. There is a view that full cost pricing is not feasible in developing countries as a tariff under full cost pricing is not affordable to people in developing countries. However, the result shows that the percentages of 30 m^3 tariff in ordinary workers' monthly income of Fukuoka City (1950's) and MM (MWCI) are 2.9 % and 3.3 %, respectively. Thus, even in developing countries, full cost pricing can be feasible if their water supply system is operated properly.

3. As for water leakage and illegal connections, the result shows that systematic replacement of old water mains to new water mains is effective in improving a RWR, rather than conducting patchwork measures such as detecting leakages and illegal connections point by point.
4. As for the view of imbalanced concession fee allocation in the process of MWSS, privatization is one of the major reasons of MWSI management failure. The result of the simulation shows that even though concession fee allocation is modified to proper ratio (MWSI: 60 %), MWSI fails its management if it does not conduct substantive measures against leakage and illegal connections, while MWCI still achieves surplus with the 40 % share of the concession fee. Therefore, the simulation implies that the major reason of the MWSI management failure is not the allocation of the concession fee but the mismanagement of improving the RWR. Also, the simulation implies that improvement of RWR is a common target both for privatized water utilities and public water utilities; thus, it can be said that the important thing for water supply system development/management is not a form of management (private/public) but an effective improvement of RWR.

In technical cooperation to developing countries, supporting their technical revel improvement by introducing current Japanese technology is one of the important tools. However, it is also important to transfer Japan's experience when its economy was current developing country level in order to show that, with proper development/management, it is feasible for developing countries to achieve self-sustainable development of infrastructure.

19.2.2 Case Study of Environmental Symbiosis Project of Kyushu University New Campus Transfer Program: Water Cycle System Preservation and Biodiversity Preservation Project

Kyushu University started a transfer program to Ito Campus, a new campus which would span the Motooka/Kuwabara districts at Nishi area in Fukuoka City, Itoshima City (former Maebaru City, former Shimamachi), under the concepts of "Building a university which transforms autonomously according to the changing times and an open one which retains energy" and "Creating headquarters of researches and education fit for such university", and research started in 1993.

The background of this transfer program were (1): Separated campuses were erecting barriers in smooth cooperation of compulsory education and majored field education/graduate school education, implementing joint research, etc. (2) The facility had become too old and space limitations meant advanced education/research, and diversification could not be accommodated. The campus did not have enough green area to be balanced. (3) The Hakozaki Campus was located in the extended approach zone for Fukuoka Airport, and aircraft noise was causing significant problems in education and research. There was concern that another accident may occur. (4) It was quite difficult to redevelop the facility into a higher

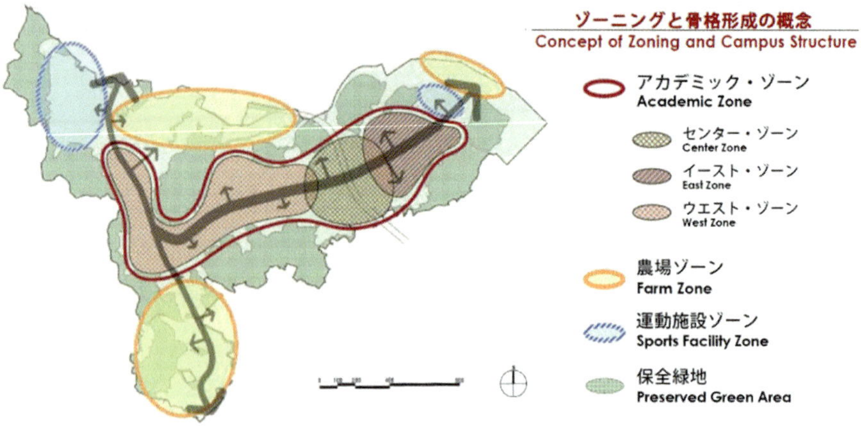

Fig. 19.4 Concept of Zoning and Campus Structure, New Campus Master plan 2001

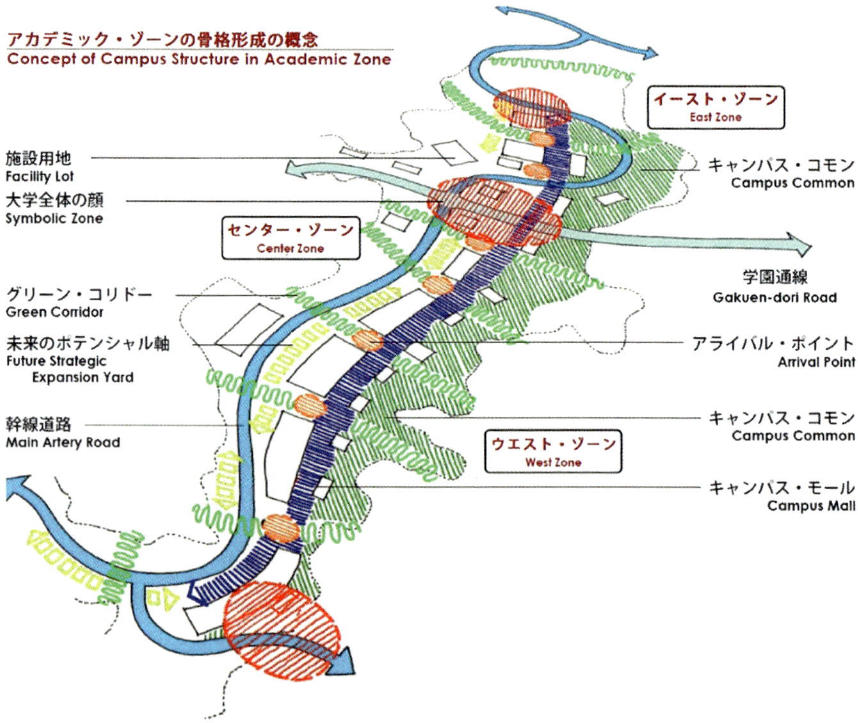

Fig. 19.5 Concept of Campus Structure in Academic Zone, New Campus Master plan 2001

and integrated one in the Hakozaki area due to various factors such as height regulations in the Civil Aeronautics Act.

The new campus construction program was given praise in the following ways: (1) It actively promoted symbiosis with the environment in a large-scale project by garnering wisdom in and out of the university (Figs. 19.4 and 19.5). (2) It was worth the attention not only as an innovative project that could become a model for large-scale projects but also in showing that a frontier case could be accomplished in future engineering works. This program by Kyushu University and Fukuoka City Land Development Corporation won the Environmental Award of the Japan Society of Civil Engineers in May 2002.

Outline of Water Cycle System Preservation Project

Preserving Water Cycle System

A considerable number of wells for living, business, and agriculture exist around the area of new campus construction, including many drawing water from the hills where Ito Campus is to be constructed. Therefore, prevent the drying up of wells due to a diminished level of underground water or well usage problems such as high salinity, adequate consideration needed to be taken into for underground water preservation in the applicable area. Kyushu University presents the following basic guideline in "Environmental Impact Assessment" concerning underground water preservation:

1. Preservation of rainwater storage and penetration capability (green zones are aggressively established; a rainwater penetration facility is installed; and a storage and penetration facility is installed underground)
2. Reducing use of underground water (main water source on campus is waterworks; introduction of recycled water use system; enlightenment activity to increase awareness of water saving)
3. Conducting underground water monitoring (monitoring underground water level, water quality)

Water Cycling System Preservation Development Plan

The "Water Cycling System Preservation Development Plan" was made in July 2004, aiming to identify the details and amount of actual measures. This is quite an advanced measure, unrivaled in Japan. For example, the underground water pres ervation goal is recovery of a decreased amount (estimated) of rainwater penetration after development by rainwater penetration facility, etc. The following penetration facilities are introduced under this plan:

1. Water penetrating pavement: Water directly penetrates from the land surface to underground.
2. Infiltration trench: Water penetrates to underground through infiltration holes of drainpipes.
3. Infiltration tank: Water penetrates from the bottom (rubble) to underground.
4. Drain ditch: Water penetrates from the sides and bottom to underground.
5. Rainwater storage and penetration facility: Rainwater is stored and penetrates to underground.

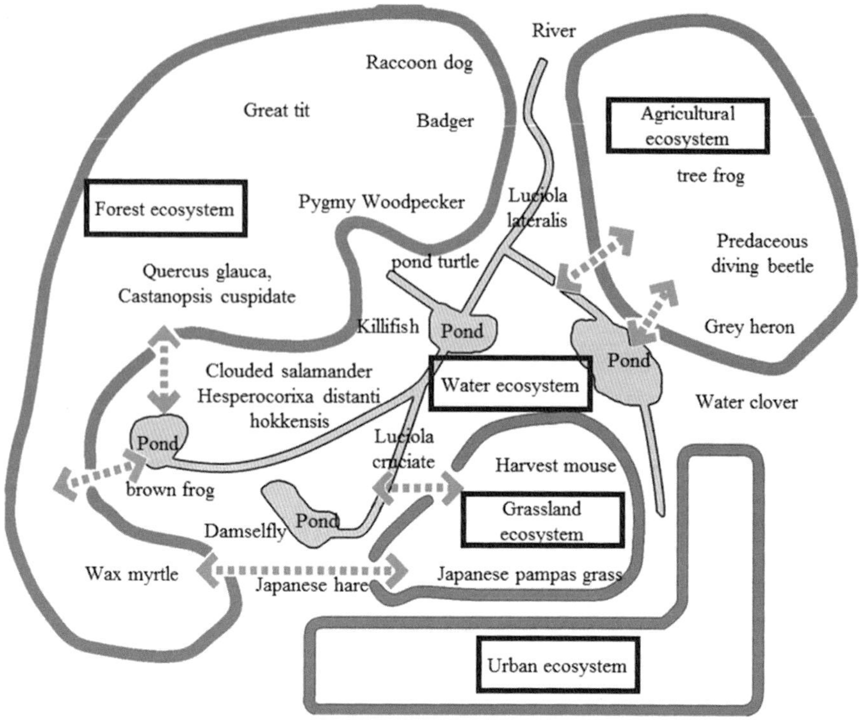

Fig. 19.6 Concept of Biodiversity Master plan 2001

The effect of the penetration facility is observed in order to be fully aware of an effective introduction method for a penetration facility to buildings which will be constructed in the future. Penetration facilities are inconspicuous. However, the whole new campus will be equipped with permeable drainpipes stretching over 28 km, over 2,000 permeable water collecting tanks, and over ten underground storage and penetration facilities according to the plan. It will cost more than conventional construction, but we think the facilities are indispensable for Kyushu University to collaborate with the local community and develop together with it.

Outline of Biodiversity Preservation Project

Objective of the Project

At the construction site of Kyushu University New Campus, a biodiversity preservation project has been in progress aiming for "no species to be lost" and "no reduction in the forest area."

Scope Management: "Biodiversity Preservation Zone"

The marshy land to the west of Sayanokami Spring, one of the valuable water sources in the area, is designated as a "Biodiversity Preservation Zone," preserving *satoyama* woodland, waterfront, and marshland environment in consideration for continuity of the ecosystem. In the preservation green zone including this "Biodiversity Preservation Zone" (Fig. 19.6) (Concept of Biodiversity Master plan 2001), rare species such as *Marsilea crenata Presl/Water clover* and *Hynobius nebulosus/ Clouded salamander* have been preserved.

Issues of the Project

Forest management in the hills became neglected after the land was purchased by Fukuoka City Land Development Corporation from the local residents for the Kyushu University New Campus. As a result, *moso* bamboo gradually proliferated. The *moso* bamboo area can expand by almost 10 m in a single year, supplanting other vegetation with its strong reproductive force. The whole of the Motooka hills will be covered with bamboo if no measures are taken, and what is worse, they say the *moso* bamboo blooms all at once and then dies off once every 100 years, becoming a factor in landslide disaster.

Activities and Solutions of the Project "Green Zone Preservation Volunteer Activity"

In 2001, at the request of the Kyushu University Administrative Bureau, bamboo trimming was conducted on holy days with cooperation of university faculty members, students, and local residents. Trimming was conducted under the instruction of Professor Shigeru Ogawa, Professor Tetsukazu Yahara, and Associate Professor Takao Setsu, joined by families of faculty and office members of the university. The bamboo shoot harvest was favorably received.

Since then, environmental preservation activities and exchange with local community have been actively conducted mainly consisting of Fukuoka Green Helper Volunteers, research on living things by citizens, and NPO Corporation Kankyo Sozosha, whose activities are highly evaluated by the local community and people involved.

19.3 Practical Education by Case Method Materials of Environmental Project Management

We tried making teaching materials by case method which is a business administration education method, for environmental issues as practical education in environmental project planning. One of the cases we used was environmental projects, "water cycle system preservation and biodiversity preservation project" in the Kyushu University campus moving project introduced in Sect. 19.2. The case method is an education method for the students to come up with solutions to actual problem/issues by themselves. Furthermore, we added the project management method when we made the case into a hypothetical project and used group work method as a virtual project team. In case method education, students understand the related information through prior reading assignments. They discuss problem analysis and solution in class lecture time and find a solution. Therefore, case method education transforms students from passive to active, and they can gain experience closer to reality. Section 19.3 introduces question items of this case method.

19.3.1 Case Method of Environmental Symbiosis and Water Cycle System Preservation, Kyushu University New Campus Transfer Program

Case number: 2012EPM-01, Version(A)17/7/2012

Consideration for the environment along with the Ito Campus transfer project (Environmental Assessment) (A)

[Outline/Background]

*Read the separate reference materials very carefully and make sure of the contents.

Kyushu University is in the process of an integrated transfer project of Ito Campus, the largest campus of Kyushu University where approximately 12,000 students, faculty, and staff will assemble as of 2012.

Ito Campus lies on some hills where ample natural beauty was left at the time of construction, as well as a valuable underground water source area used by the local residents for daily living requirements, business, and agriculture. Therefore, it was necessary to prevent the drying up of wells due to a diminished level of underground water or well usage problems such as high salinity.

Also, Sayanokami Spring rising from the marshy land in the center of the campus was a valuable water source for agriculture, and the existence of important creatures such as *Hynobius nebulosus/Clouded salamander* and *Asellus hilgendorfii/Hesperocorixa distanti hokkensis* was confirmed.

Therefore, Kyushu University decided to develop several projects and conduct environmental assessment in order to proceed with the transfer project in a spirit of full consideration for the environment.

[Assignment 1] Discuss what projects need to be promoted based on the current issues.

[Assignment 2] *Group

Follow the PDM items to suggest a project including project name, project execution agency and names of members, project time, project schedule, purpose of the project, aim, outside condition and prerequisite condition [risk management analysis], solution and method (strategy), input, and project activity.

1. Team building: Introduce yourselves simply to your group (nickname, tell them what your strong points are) (i.e., what contribution you can make to group work).
2. Begin group work after deciding on each role and agreement with time schedule by everyone (Confirm work contents, manage your time to come up with results by the final work time (5th period of July 24)).
3. Use the analysis methods you have learned so far. Examine the following one to seven and include them in your presentation.

   The presentation should have the names of participants in your group, background of the project, raising the question, goal setting, extracting issues, risk management, drawing up and deciding on solutions, evaluation/selection/announcement, implementation plan drafting, and input.

   1. Analyze who the stakeholders (people involved) are in each project.
   2. Analyze what requests can be expected from the stakeholders to Kyushu University [specifying needs]. What are the problems in transfer?
   3. Think and analyze what the overall goal and project purpose of the project are.
   4. Discuss and decide which environmental categories must be examined [confirming approach].
   5. Analyze and think how much impact can be expected in the topic 2.
   6. Decide how the measures are conducted [goal analysis] [output] [activities].
   7. What methods should be used to confirm preservation condition?

4. Use PowerPoint for the presentation, and each group shall prepare posters.

19.4 Conclusion

This chapter introduced the basics of project management and case study teaching material for environmental projects. Empirical knowledge and basic knowledge of management, in addition to the grounding management, are necessary to develop strategic management. Projects exist in the lives of students, but it is probably students without empirical knowledge at all can actually feel the importance of project management when students will go out into the world. We did not refer to all of the lecture materials here. We hope students enjoy to study "project management" by case method. We try to sum up another case study methodologies for project finance next version due to the limitation of pages. We would like to include these items in a future publication.

References

Foundation for Advanced Studies on International Development (2008) PCM-management Tool for Development Assistance (Participatory Planning), 7th edn. Foundation for Advanced Studies on International Development, Tokyo

Project Management Institute (2013) Project management body of knowledge (PMBOK®GUIDE), 5th edn. Project Management Institute, Philadelphia

Tsukahara K, Takigawa N, Koriyama K, Fujii T (2008) Comparative study on water supply system management focusing on a non-revenue water ratio in post privatization Metropolitan Manila and 1950s, vol 2. Urban Policy Studies, Fukuoka City, pp 85–97

Related Articles

Angus Maddison (1995) Monitoring the world economy, 1820–1992

Asian Development Bank (2004a) Project completion report on the Angat water supply optimization project. Asian Development Bank, Manila

Asian Development Bank (2004b) Project completion report on the Umiray-Angat Transbasin Project Asian Development Bank, Manila

Fukuoka Prefecture (1961) Fukuoka-ken Toukei-Nenkan (in Japanese)

Fukuoka City Water Works Bureau (1976) Fukuoka-shi Suidou Gojunen shi (in Japanese)

Fukuoka City Water Works Bureau Fukuoka-shi Suidou Toukei Nenkan (1950–1955) (in Japanese)

Nippon Jogesuidou Sekkei CO., LTD., Tohmatsu & CO (1996) Study on water supply and sewerage master plan of Metro Manila in the Republic of the Philippines. Main Report

Nippon Jogesuidou Sekkei CO., LTD., Tohmatsu & CO (1996) Study on water supply and sewerage master plan of Metro Manila in the Republic of the Philippines. Supporting Report

Overseas Research Department Japan External Trade Organization (2006) The 16th survey of investment-related cost comparison in major cities and regions in Asia

Urban Research Center of Fukuoka City (2007) Ajia chiikini okeru Tsoshibu Hinkonsouno Mizukyoukyuni Kansuru Kenkyu (in Japanese)

Chapter 20
Foundation for the Establishment of Urban and Environmental Plans in Asia and the Required Practical Education

Hiroki Nakamura and Koichiro Aitani

Abstract The basic ideas needed to establish urban and environmental plans in Asia, and the practical education required to accomplish this, are introduced in this chapter. The quality of the urban environment has declined in many large cities in Asia due to rapid urbanization, economic development, and motorization. Therefore, it is indispensable to make urban and environmental plans to secure healthy and cultural urban life and functional urban activities. At the same time, it is important to maintain a healthy harmony with the agricultural, forestry, and fishing industries. This must be achieved by effectively distributing the limited land resources within the city and by appropriately allocating land for construction, infrastructural facilities, green land, and natural environment. Accordingly, this chapter will introduce the foundations of urban and environmental planning in Asia. It will be explained based on the idea of urban planning in Japan, with practical case studies from Kyushu University.

Keywords Urban plan • Master plan • Environmental plan • Urban facility • Residential facilities • Transportation facility • Waste treatment facility

20.1 Cities in Asia

In many large cities in Asia, the urban environment has been in decline for some time now due to rapid urbanization, economic development, and motorization. While many countries are faced with rigorous resource restrictions, supply cannot keep pace with demand for infrastructural facilities in Asian cities. Take transportation facilities for example; underdeveloped road networks and public transportation have often paralyzed urban functionality, even while the population is increasing rapidly. Urban transportation serves as the arteries of a city and has a major impact on urban structure, as well as directly affecting the complicated

H. Nakamura (✉)
Department of Urban and Environmental Engineering, Kyushu University, Fukuoka, Japan
e-mail: nakamura@doc.kyushu-u.ac.jp

K. Aitani
Department of Architecture, Texas A&M University, College Station, TX, USA

T. Shimaoka et al. (eds.), *Basic Studies in Environmental Knowledge, Technology, Evaluation, and Strategy*, DOI 10.1007/978-4-431-55819-4_20

movement in and out of the city. Without a smooth flux of people and goods throughout the city, there will be serious impacts on the lives of residents. Infrastructure related to water services (water and sewage) and energy (electricity and gas) are as critical as transportation. Facilities equivalent to the veins of a city are those related to removal and treatment of waste and wastewater. It is indispensable to make an urban and environmental plan by which to secure healthy and cultural urban life and functional urban activity. This can only be achieved by effectively distributing, and appropriately placing the required facilities and equipment, within the limited land available in rapidly changing Asian cities.

At present, many large Asian cities have a monocentric layout. The central part includes provisions for various functions (e.g., commercial facilities, administrative agencies, residential facilities) that are directly linked to transportation facilities (Morichi and Acharya 2013). For this reason, cities keep expanding, after which even more roads must be built to respond to the urban expansion (Fig. 20.1).

Such monocentric urban layouts have some advantages for efficiently energizing the economy. However, it also has a negative side: being unable to meet the fluid demands for development of infrastructure in many of the most rapidly growing cities in Asia. There have been numerous discussions concerning how cities should be laid out. In summary, these types include the monocentric, polycentric, and network, as seen in Fig. 20.2. A consensus is being reached that the polycentric type is the ideal layout for a city (Morichi and Acharya 2013). This point needs to be fully acknowledged while establishing urban and environmental planning in Asia.

Here, we will discuss the appropriate size of a city and its ideal layout according to the simple analysis framework by Batten (1995), as is shown below.

First, the formulae are defined:

$$X = a\left\{b(P+G)^2 - P^3 - G^3\right\} \tag{20.1}$$

$$P = z\left(P^* + D\right) \tag{20.2}$$

$$G = z' G^* \tag{20.3}$$

where X is economic activity level in the city, P^* is the population active in the urban economy, D is outside customers, G^* is the stock of facilities (infrastructure)

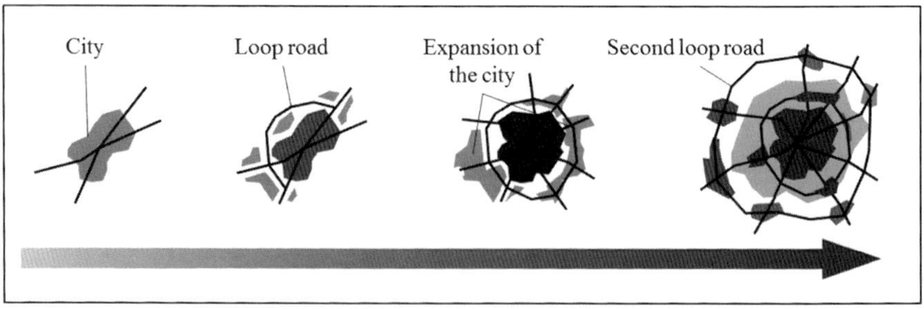

Fig. 20.1 Image of an expanding city (Source: Based on GTZ 2004)

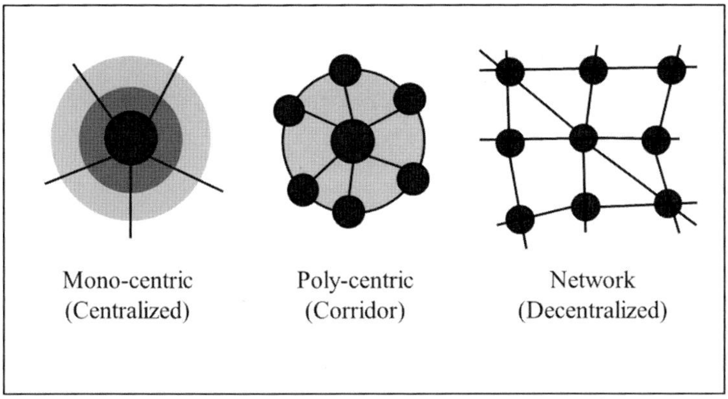

Fig. 20.2 Types of urban structure (Sources: Morichi and Acharya 2013; Batten 1995)

in the city, "*a*" is the technology efficiency index, "*b*" is the economic activity scale index, "*z*" is the human capability index, and "z'" is the quality index of facilities (infrastructure). Formula 20.1 shows the urban production function, wherein *D* indicates the role of the city as an international market.

When economic activity reaches its maximum in the city, the marginal productivity of *P* and *G* reaches zero. Reducing the partial differential simultaneous equations will give us the ideal condition:

$$P = G = 4b/3 \tag{20.4}$$

Next, because marginal productivity and average productivity are equal at the optimal size, we have:

$$\frac{P^3 + G^3}{(P + G)^2} = \frac{b}{2} \tag{20.5}$$

Here, let us consider primitive urban economic activity without the stock of facilities (infrastructure) in the city. When $G = 0$, (20.1) becomes:

$$X = a\{bP^2 - P^3\} \tag{20.6}$$

which is shown in Fig. 20.3, and (20.5) becomes:

$$P = \frac{b}{2} \tag{20.7}$$

Proceeding to a discussion of city types shown in the figure, given this simplified urban economic activity, when the economic activity of a city reaches its maximum, the slope is 0, that is:

Fig. 20.3 Relationship between the size and output (production function) of a city (Source: Based on Batten 1995)

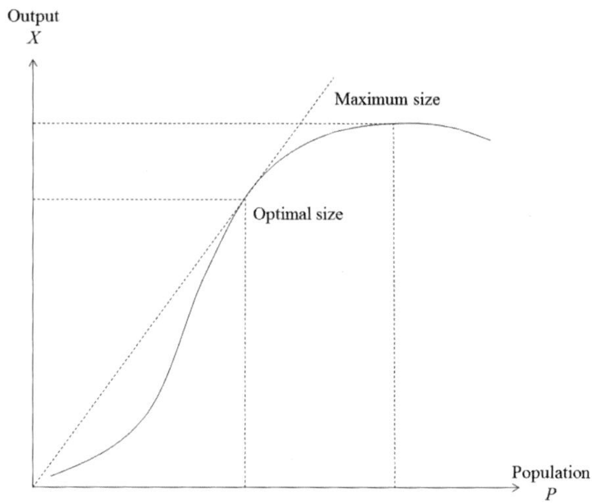

$$P = \frac{2b}{3} \tag{20.8}$$

Therefore, the relationship between urban economic activity and its population size is described by the following rules:

1. Gradual increase in productivity along with expanding size when $P < b/2$
2. Gradual decrease in productivity along with expanding size when $b/2 < P < 2b/3$
3. Decrease in productive activity along with expansion of its size when $2b/3 < P$

As is seen here, we can gain interesting insights for discussion concerning the appropriate size of a city, and how a city should be laid out, when we come up with primitive urban economic activity that could be considered in reality.

1. When $P < 2/b$, network cities with nodes of different sizes can have higher productivity than network cities with nodes of the same size.
2. When $P = 2b/3$, monocentric and polycentric cities have about the same level of productivity, if they are about the same size.
3. When $2b/3 < P$, network cities can have higher productivity than monocentric cities, and the productivity of a network city with nodes of the same size is greater than that of a city with nodes of different sizes.

However, this analysis framework by Batten (1995) is an extremely simplified example, and in reality, it is necessary to think practically about the ideal layout of each individual city. That is to say, each urban and environmental plan should consider a variety of factors, including the existing stock of urban facilities (infrastructure) and the historic background of development. Fundamental concepts for putting urban and environmental plans into practice are provided in the "Guideline for Urban Planning Procedures" by the Ministry of Land, Infrastructure, Transport and Tourism, Japan (2008).

20.2 Foundation for Establishing Urban and Environmental Plans

20.2.1 Master Plan

The foundation of urban and environmental plans should be the goal of securing healthy and cultural urban life and functional urban activities. This can be accomplished by efficiently distributing the limited land resources within the city, and appropriately allocating land for construction, for infrastructural facilities, for green land, and to retain as natural environment. At the same time, it is important to maintain healthy harmony with the agricultural, forestry, and fishing industries. It takes time to achieve these aims, and the effort needs to be approached with a long-term view. It is also important to explain successfully the necessity and validity of the urban and environmental plans determined to be best. Since these two (urban and environmental plans) form an integrated whole, they always need to be examined from the viewpoint of comprehensiveness and unity, based on their relation with projections of how the city should be in the future. It is quite important to identify the future layout of the city from a long-term perspective, in a form that is easy for residents to understand. With this goes the need to clarify the path to its realization, and this role is expected of a master plan. The following list constitutes the major content of a master plan:

1. The future layout of the city, based on a long-term viewpoint, is identified in consideration of the trend of urban development, the population in the area affected by the urban plan, and current and future prospects of industries (projections into the future), while a major path to their realization is identified.
2. The future layout (roughly 20 years later) of the city is made, and the basic direction of urban planning is determined. However, regarding the scale and areas of the zones designated for urbanization within roughly 10 years, it is desirable to have a 10-year projection. In addition, it is desirable to place priorities and set a goal of development, within roughly 10 years, for urban facilities and development projects in areas designated for urbanization.
3. The master plan should not be a mere lengthy description concerning the individual urban plan, but should indicate what city is going to be built under the plan. It is also desirable to show the rough location and size of future land uses, urban facilities, and urban development projects. In addition, it is desirable to show the contents of the master plan using media that are easy to understand. Topographical maps or image figures can be used in such cases.
4. In order to deal with situations that change after the master plan is established, it is desirable to take the following measures and provide for follow-ups. First,

make the description flexible, for example, it should be limited to items whose future layout, at the time of determination, is available to a certain extent, and other items should be added as necessary, at such time as their reasonably clear layout becomes available. Second, when partial modification of the plan must be conducted, it is done in a timely fashion and without delay.

5. Upon establishing a master plan, it is desirable to visualize the current situation and future layout (of the project areas) in relation to adjacent urban planning zones in consideration in order to secure a wider point of view. A case could be made, for example, for establishing a wider-area (regional) master plan, covering multiple urban planning areas, based on the master plans established for each individual urban planning area.
6. Specific urban plans have an obligation to comply with the master plan. This means specific plans should not contradict the future layout of the city targeted by the master plan nor the pathway by which the plan is to be actualized. This does not mean that the description of each individual urban plan must be in the master plan.
7. Upon establishing a master plan, it is desirable to pay full attention to the impact of the items contained in it, on the living conditions, functions, and natural environment of the city.

20.2.2 Urban/Environmental Facilities

Urban and environmental facilities are indispensable to support smooth urban activity, improve convenience for urban residents, and secure a satisfactory urban environment. It is useful, when deciding on what urban and environmental facilities to provide in urban and environmental plans, to consider the processes listed below.

Clarification of zones necessary for development at the planning stage: By clarifying the zones necessary for developing urban facilities, it becomes possible to expand systematic development over the long term. This should enable smooth, steady development of urban facilities.

Adjustment of land use plans between individual cities and environmental facilities: By adjusting land use plans within cities, and among individual cities and environmental facilities, it becomes possible to proceed with comprehensive and integrated development of the cities.

Promotion of consensus among residents: The size and location of the facilities needed in future cities are greatly influenced by the residents and other stakeholders. It is best to establish consensus building in the local community through open procedures and generous sharing of information.

The following are some ideas for practical planning of individual urban or environmental facilities, based on the criteria just described:

20.2.3 Residential Infrastructure

The purpose of residential infrastructure of a city is to secure adequate shelter for the human population in the city. When inadequate, it is increased by constructing residences and associated amenities that provide satisfactory living conditions. To accomplish this, it may be necessary to provide new facilities to increase the convenience of residents in particular areas, and these provisions should be based on a comprehensive land use plan for the city.

Plans for providing urban residential infrastructure should utilize appropriate zones and scale, so that fine residential areas result. These activities should be informed, first, by consideration of a land use plan, hopefully provided within the context of a master plan, and, second, by the situations in the surrounding urban areas. When doing so, it is desirable to consider how to provide some hours of sunlight, good lighting, ventilation, privacy, and healthful views (beautiful scenery) for residents, as well as the best location of roads, parks, and city squares consistent with the topography. Moreover, all of this needs to be done considering harmony with the neighboring urban area, in order to secure a satisfactory living condition when locating residential facilities. There will be restrictions related to floor area ratio and building coverage ratio. It is desirable to set such limits with consideration for the major uses of the district and adjacent districts, the existing district plan, and conditions in neighboring urban areas.

20.2.4 Transportation Infrastructure

Individual transportation facilities should be established to provide a comprehensive, integrated transportation system within which each transportation mode (public transportation, vehicles, bicycles, pedestrians) plays a properly allocated role in the whole. This must be done from the viewpoint of user convenience and transportation efficiency and to preserve a satisfactory urban environment with consideration for global environmental issues. Since the transportation infrastructure forms the framework of the future city, it is desirable to set goals (e.g., higher-density urban areas along public transportation routes) for urban planning that include the transportation system in the master plan. In addition, it is desirable to examine the organization of the transportation system as being integrated with the land use, in order to achieve these goals.

First, upon establishing an urban road plan to achieve the nominal (target) goals of the city, it is desirable to consider the details (e.g., layout, number of vehicle lanes based on the planned volume of traffic). In addition, space for pedestrians, bicycles, and public transportation (e.g., streetcars and buses) needs to be considered. The various functions of roads must be provided according to their roles. Moreover, the location of radial roads and loop roads must be carefully examined so that the full potential of other roads can be fully exploited. These examinations are generally conducted based on urban traffic surveys. In particular, in existing urban areas where space is limited, it is also desirable to determine the best locations for transport facilities that consider adjustments in traffic demand and future use of transportation facilities to be developed. It is also important to fully examine the targets for service standards and the number of individual facilities that will be needed.

In determining road plans in cities, it is desirable to form the new urban planning road network, by adequately combining existing expressways, main roads, utility roads in districts, and special service roads, in full cooperation with other urban facilities. Needless to say, this must be done with consideration for consistency with the wider area network and land use, so that the location of roads becomes the framework of the city. Furthermore, the plan should be determined based on natural conditions, including topography and geological features, as well as social conditions (e.g., the form of the urban area, current land use, and the natural and historical sites or districts that will need to be preserved).

20.2.5 Waste Treatment Infrastructure

Regarding waste treatment facilities, when decisions have been made that result in a comprehensive, integrated urban and environmental plan, adjustments in consideration of other urban and environmental plans, and public consensus, can be arranged among interested parties, enabling smoother development. In particular, the location of industrial waste treatment facilities tends to be problematic. The role of location planning is expected to be involved when urban and environmental plans include industrial waste treatment facilities, most of which are large scale. Therefore, it is essential to provide a viewpoint that favors integrated, comprehensive urban and environmental plans.

Moreover, a final treatment facility is expected to have a permanent nature, considering also the appropriate use of its site after demolition. It is thus subject to major decisions in urban and environmental plans.

20.2.6 Water Treatment Infrastructure

Water drainage and treatment facilities are indispensable for effective urban activity since they determine many factors related to improvement of living conditions (e.g., effective removal of domestic waste, storm drainage, road use in bad weather) and for preservation of water quality. Therefore, it should have a prominent place in urban plans. Plans related to water and sewage services require comprehensive consideration of the natural condition of the land, trends in land use, the current level of development waterways, as well as their projected futures. The water emission area, treatment facilities, pump stations, and main pipes should be integrally and comprehensively determined so that urban functions are assured and a satisfactory urban environment can be formed and maintained.

Associated with water drainage services are the essential functions provided by rivers in cities, especially in highly populated areas, in addition to their basic role in flood control. Therefore, rivers play important roles in promoting the development of healthy, comfortable, and more interesting cities. These functions include:

1. Open space that refreshes and comforts urban residents and at the same time provides space where various animals and plants may live
2. Potential for creation of attractive urban scenes that connect with more natural areas along the river
3. Potential for creation of pleasant, energizing waterfront areas usable by the public for gatherings and events
4. Potential for use in disaster prevention, as an evacuation site, as part of evacuation routes, for access to emergency transport by boat, as a firebreak belt, and as a major source of fire-fighting water in the event of a major disaster

Rivers improve safety in urban areas when improvement work is done. Furthermore, the facilities that provide the specific functions listed above are closely connected with land use functions and urban facilities in the surrounding area. Therefore, they should have a prominent place in urban and environmental plans. Particularly in areas designated for urbanization, urban plans for rivers should be determined in the same way as roads, parks, and the sewage system. It is also desirable to make decisions in harmony with other urban facilities (e.g., roads, parks) that contribute to improvement of the urban environment, by considering current and future land use in surrounding areas. When the urban plans of rivers are determined, local visions for community improvement that involve rivers should be examined in relation to the need for access to the river. It is also helpful to coordinate with park/green land and commercial plans for land use and to accommodate restrictions and guidelines concerning uses and height of construction along the river, in order to create and maintain beautiful scenery. It is desirable to establish urban river plans that consider relationships with urban facilities that include roads and parks, with areas designated for urban development, if necessary.

20.3 Practical Education for Establishing Urban and Environmental Plans in Asia

Kyushu University provides the "Exercise: East Asia Project Study" program for the study of different, but converging, fields of study. Students get practice at setting up and solving problems in a group lodging style. They are able to experience the significance of establishing urban and environmental plans and processes via simulation. A joint program ("Exercise: East Asia Project Study" with the "Sustainable Design Camp") has been conducted since 2011, as part of an educational program for building a sustainable urban construction system in Asian cities by the Department of Architecture and Urban Design, Graduate School of Human-Environment Studies, at Kyushu University. In addition to the educational problems focused on architecture and urban design, this exercise looks at environmental problems of wider scope, including waste treatment and public sanitation, among others. The task assigned to the students is to explore approaches for strategically solving urban environmental problems with a wide scope and make suggestions to realize the creation of sustainable cities where convenience, comfort, and safety for the residents are achieved while still having consideration for the environment.

The first event in FY 2011 was held in the Warasamra area, suburb of the commercial city of Colombo, in Sri Lanka, in cooperation with the University of Moratuwa. The second event (in FY 2012) was held in Ho Chi Minh City, Viet Nam, in cooperation with Van Lang University, Ho Chi Minh City University of Technology, and Ho Chi Minh City University of Architecture. Nepal was chosen as the study area in FY 2013, and further economic development requiring solution of environmental problems is expected there. The exercise was jointly hosted from July 26 to August 06, by the local Tribhuvan University and Chinese Culture University (Taiwan) in cooperation with the Nepal office of UN-Habitat.

During the exercise, the following topics were presented after being assigned to students for group work: "(1) urban and environmental master plan, (2) waste treatment, (3) water supply and wastewater treatment, (4) core design (complex facilities in crucial districts in a city), and (5) residence plan." Groups were set up for each topic, and the participating students made suggestions for plans and designs as problem solutions per group under the instruction of professors. The specific process in the exercise was divided into the following eight stages:

1. Sharing basic information: Clearly grasp the purpose of the exercise and its position (context). In particular, identify the economic and social positioning of the target country and city, and consider why the study area was chosen.
2. Site survey (study area and related facilities): In order to grasp the outline of the study area, actually visit the study area and related facilities in the vicinity for research, on foot.
3. Collection of local information and interview of residents: Properly collate the issues of urban planning and environmental problems as well as information concerning policies for the measures to be taken. This should be done by attending administration hearings or lectures at universities, as well as by reference to existing information about the study area. In addition, ask citizens to present opinions concerning the current situation and their ideas regarding what should be done.
4. Gaining a first impression: Before obtaining and analyzing detailed information of the study area, students walk around the study area without preoccupation and observe the features of the study area intuitively.
5. Analysis of the site areas: Study and analyze the topography, geography, history, nature, and environmental quality to extract "Strengths" and "Weaknesses" in

the study area, and then analyze and grasp what the outside "Opportunities" and "Threats" are (SWOT analysis). "Problems" and "potentials" in both the inner and outside environment are extracted.

6. Future vision of the district: Think about how the problems in the study area can be solved and what an attractive place it could become. Then search for a way that the future vision could be established and what strategic design method will be available. Discuss the vision of what it should become in the future. Through discussion, determine how the "problems" should be overcome while a specific strategy is examined in order that the "potentiality" can survive.
7. Presenting a specific design and plan for an effective site: Focus on the specific location or facility that seems to be the most effective and propose a specific plan or design. Consider the context in order to show actual prototype examples of the future vision and strategic design method for the discussed study area.
8. Scenario proposal in order to implement the plan: Refer to the legal, financial, and economic basis, as well as the control/management system after completion. Put the final plan/design into practice so that the existence of a realistic and feasible scenario, as opposed to a fictional one, can be shown. It is hoped that the scenario in the timeline can be presented in stages: short term, middle term, and long term.

Kathmandu, Nepal, which was the subject of exercise in 2013, experienced rapid urban expansion from 1967 to 2011 as is shown in Fig. 20.4. Loop roads were constructed as if accommodating the expansion of the city, resembling the previously shown Fig. 20.1. On the other hand, construction of a second loop road is planned as is shown in Fig. 20.4. Therefore, for this exercise, Kirtipur was chosen as the subject. This is a smaller city that once flourished in Kathmandu Valley where the second loop road crosses and urbanization is in progress and has the potential to become an important satellite city (Fig. 20.5).

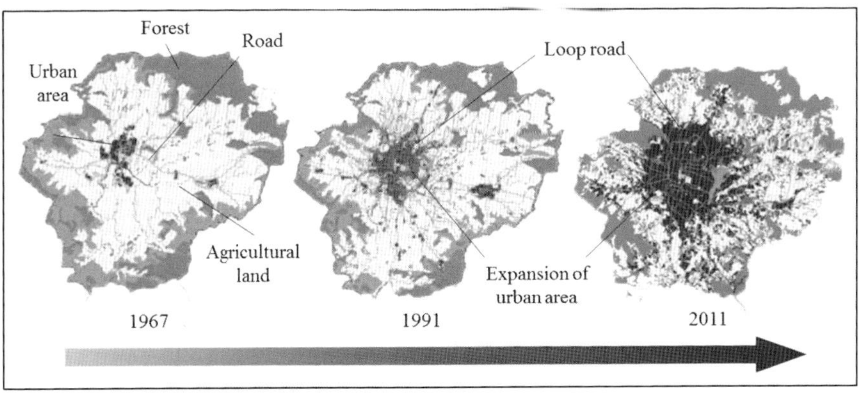

Fig. 20.4 Process of urban development of Kathmandu

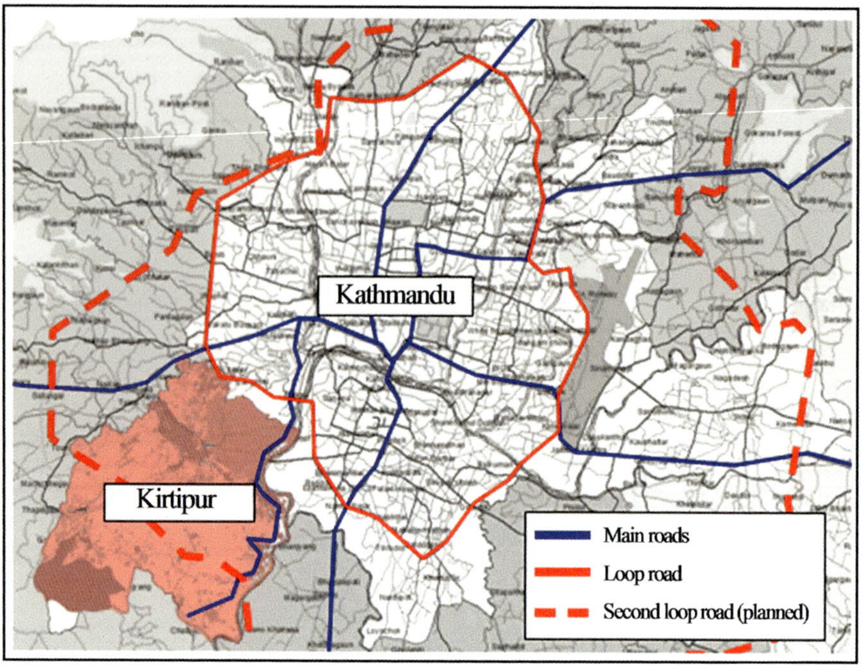

Fig. 20.5 Main roads and location of subject area of the exercise in Kathmandu

Table 20.1 Partial SWOT analysis done of the subject area done by students

Item	Detail
Strength	Historical townscape, rich cultural heritage, green-rich nature
Weakness	Inferior quality infrastructure, poor habitat environment, underdeveloped traffic network and road environment
Opportunity	Development as a tourist site, population increase
Threat	Rapid urbanization and environmental pollution, disaster risk

In the actual exercise, the participating students carried out a SWOT analysis in the subject area partially shown in Table 20.1, using the previously mentioned eight-stage process. This became the basis for the master plan at every phase partially shown in Table 20.2. Based on the master plan (the land use plan shown in Fig. 20.6), suggestions were made on the designs of temples, public facilities, and city square shown in Fig. 20.7. Suggestions from the students were compiled and published in a book (Shimaoka et al. 2014). The exercise is practical education to understand the foundation for the establishment of urban and environmental plans in Asia.

Table 20.2 Partial urban and environmental plan at each phase planned by students

Phase	Details
2013–2020	Detailed traffic survey, road development, installing bus stops, development of water supply base and open space, road development for festivals and tourism, house improvement, etc.
2020–2025	Development of roads/paving, organizing bus route network, installing disaster control center, maintenance/management of tourist sites, repair of temples, etc.
2025–2030	Development of festival route, installing bus lanes on the road, development of local disaster prevention bases in populous areas, improvement of scenery and waste management at tourist sites, development of tourist facilities such as restaurants, souvenir shops, and rest areas, etc.

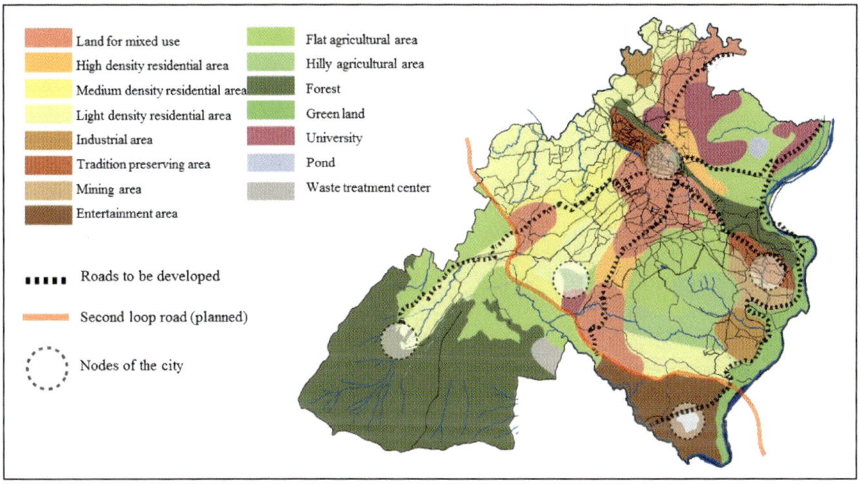

Fig. 20.6 Land use plan for the subject area, made by students

Fig. 20.7 Temples and public facilities in the subject area, designed by students

References

Batten DF (1995) Network cities: creative urban agglomerations for the 21st century. Urban Stud 32(2):313–327

GTZ (ed) (2004) Land use planning and urban transport. http://www.sutp.org/files/contents/documents/resources/A_Sourcebook/SB2_Land-Use-Planning-and-Demand-Management/GIZ_SUTP_SB2a-Land-use-Planning-and-Urban-Transport_EN.pdf

Ministry of Land, Infrastructure, Transport and Tourism, Japan (2008) Guideline for urban planning procedures, 6th edn. http://www.mlit.go.jp/common/001091846.pdf (in Japanese)

Morichi S, Acharya SR (eds) (2013) Transport development in Asian megacities. Springer, Berlin

Shimaoka T, Kanno T, Nakayama H, Aitani K, Nakamura H (eds) (2014) Re-birth of Kirtipur: awareness of gifted nature and cultural landscape for wellbeing of citizens. Hana-Shoin Corporation, Fukuoka

Printed by Printforce, the Netherlands